HUMAN EMBRYONIC STEM CELLS

AN INTRODUCTION TO THE SCIENCE AND THERAPEUTIC POTENTIAL

ANN A. KIESSLING
HARVARD MEDICAL SCHOOL

SCOTT C. ANDERSON

JONES AND BARTLETT PUBLISHERS
Sudbury, Massachusetts
BOSTON   TORONTO   LONDON   SINGAPORE

World Headquarters
Jones and Bartlett Publishers
40 Tall Pine Drive
Sudbury, MA 01776
978-443-5000
info@jbpub.com
www.jbpub.com

Jones and Bartlett Publishers Canada
2406 Nikanna Road
Mississauga, ON L5C 2W6
CANADA

Jones and Bartlett Publishers International
Barb House, Barb Mews
London W6 7PA
UK

PRODUCTION CREDITS
Chief Executive Officer: Clayton Jones
Chief Operating Officer: Don W. Jones, Jr.
Executive V.P. & Publisher: Robert W. Holland, Jr.
V.P., Design and Production: Anne Spencer
V.P., Sales and Marketing: William Kane
V.P., Manufacturing and Inventory Control: Therese Bräuer
Executive Editor, Science: Stephen L. Weaver
Managing Editor, Science: Dean W. DeChambeau
Senior Marketing Manager: Nathan Schultz
Senior Production Editor: Louis C. Bruno, Jr.
Composition: International Typesetting and Composition
Cover and Text Design: Anne Spencer
Printing and Binding: Malloy
Cover Printing: Malloy

About the Cover: Four-cell human egg that contains three sets of chromosomes instead of two at the zygote stage; stained with fluorescent dye to visualize the DNA; one blastomere contains two nuclei instead of one. Photo by A. Kiessling.

Library of Congress Cataloging-in-Publication Data unavailable at time of printing.

Printed in the United States of America
07 06 05 04 03   10 9 8 7 6 5 4 3 2 1

To Mike and Larry, men with hemophilia; wonderful sons, husbands, and fathers who were infected with the AIDS virus and hepatitis C by tainted blood clotting factors prescribed to treat their hemophilia; men who died young, before stem cell therapy could cure them.

Brief Contents

Contents

Preface

THE SOCIAL DEBATE AND RESULTING MORATORIUM by the Bush Administration against federal funding for embryonic stem cell therapy prompted the writing of this text. That the debate took place speaks highly of our society's genuine concern for the protection of its most vulnerable members. And there is no doubt that the White House moratorium stems from a genuine belief that human life begins at fertilization. What is repeatedly missing from the debate process, however, is a thorough understanding of the biology involved. The debaters opposed to research that would destroy human embryos liken the process to abortion and murder. Those favoring this research justify the destruction of embryos for the sake of suffering persons. Neither viewpoint is accurate. And it is equally important to understand that many of the facts necessary to establish an entirely accurate viewpoint are not presently known because the necessary research has not, and cannot, be done in the United States.

The emerging field of human embryonic stem cell biomedicine crosses many disciplinary boundaries, including reproductive biology, embryology, cell biology, molecular biology, endocrinology, immunology, fetal medicine, transplantation medicine and surgery. Clearly what is needed is a single reference that provides the basic information from these multiple disciplines as it pertains to the science of human embryonic stem cells. Because the field is so new, and the social debate so heated, a reference that could be understood by both the lay public and by scientists would be especially valuable. That has been the goal of this undertaking.

Human Embryonic Stem Cells: An Introduction to the Science and Therapeutic Potential was written to meet not only the needs of science students with a basic understanding of the principles of cell biology, but also science writers, politicians, teachers, medical students, physicians, nurses, veterinarians, and biomedical scientists who may not be familiar with the various stem cell disciplines. Because of the newness of the field, we thought it important to include some historical perspective, generally in the form of scientific endeavors that advanced the field. These are presented throughout the text as sidebars. Not only will they enhance the reader's understanding of how the field has evolved, they also provide scientific details important to understanding the overall biology. Certain chapters also contain more advanced descriptions of what is known about some cell processes. This information is not essential to understanding what follows in succeeding chapters, but is designed to enrich the science for those who are interested. Also in each chapter we have attempted to identify which important scientific facts are missing, and in need of discovery.

We apologize to our many colleagues whose work was not included or adequately emphasized. We anticipate hearing from them and including their suggestions in future editions. Our thanks to our families for their patience and support during this project, to Ryan Kiessling for bringing a recent Biology graduate's view to the information presented, to Steve Weaver, Lou Bruno, Dean DeChambeau, and Anne Spencer, the competent and supportive Jones and Bartlett team, and to artist Elizabeth Morales for so skillfully rendering the artwork.

We have developed the information in the book with the view that there is no substitute for knowledge when attempting to formulate an opinion about a new, and awesome, biomedical development. Those who are firmly on one side or the other of the debate may not be swayed from their positions by the information presented. We hope, however, that those scientists, developing scientists, teachers, writers, politicians, physicians and patients struggling to fully understand the field, its therapeutic potential, and its potential pitfalls, will find after reading this book that they have a new understanding of the field of human embryonic stem cells and will be able to participate more fully in its development and its oversight. Given the broad scope of its therapeutic potential, there is a great likelihood that all readers of this text will become beneficiaries of stem cell technology at some point in their lives.

A.A.K.
S.C.A.

Foreword

LET US DREAM about the possibility of an unlimited supply of cells capable of generating any tissue of the human body. A few years ago this could have been the subject of a science fiction novel; today they are called human embryonic stem cells.

Can they replace lost cells?
Can whole human tissues be recreated?
Can they prevent age-related illnesses?
Can we find the perfect vehicle to deliver therapy?
Can we extend the lifespan of the human species?

The answers are in this book.

Since 1998, when the first report of human embryonic stem (ES) cells was published, a constant stream of articles has appeared in peer-reviewed journals and the general scientific media. The scientific community, as well as the public at large, has tried to keep up with this new, emerging area of biomedicine, but, unfortunately, a large portion of our society remains unaware of what these cells are and what implications they may have for therapeutic alternatives. A recent survey published in *Science* revealed that stem cell–based therapies will be able to cure more than 125 million people in the United States alone.

On the eve of this medical revolution our society remains divided. There are groups advocating that ES cells are the solution for many (if not all) diseases that today have no solution, while others think this is just one more approach that should be avoided because of the ethical challenges it poses. Who is right? Will we miss something important if we stop pursuing this new area of medicine? It is impossible to make a judgment without having all the information available, and that is the greatness of this book. It embraces us all, no matter how weak or strong your background in science, this book describes in an amicable way, with great simplicity and strong scientific foundations, the wonders of these cells.

With the certainty of someone that has seen the beginning, and, perhaps, could have predicted this revolution years ago, Dr. Kiessling introduces each concept at the proper time and with the essential information. In section one, we find the basic definitions of embryonic stem cells and a comparison with other, more restrictive stem cells in the body. Section two describes the role of the egg and the embryo in the formation of these cells. Here we appreciate Dr. Kiessling's deep knowledge of current and seminal literature and, surprisingly, we soon discover that what is now the "unexpected" derivation of embryonic stem cells from a preimplantation embryo is merely the consequence of a well-orchestrated cellular process that started years ago during oogenesis. Sections three and four explain in easy-to-understand detail the process of differentiation and the potential medical applications of these cells. Section five eloquently addresses the societal debates surrounding embryonic stem cells, an area of particular interest to coauthor Scott Anderson.

In summary, Dr. Kiessling and Mr. Anderson have given us a wonderful gift, a detailed narrative of human embryonic stem cells, their origin, the current state of the art, and the potential implications in basic research and human medicine. No other authors have done so. Thank you; it was time.

Jose Cibelli
Professor of Animal Biotechnology
Cellular Reprogramming Laboratory
Michigan State University

PART I

The Basics

CHAPTER 1

The Nature
of Stem Cells

All truth passes through three stages. First, it is ridiculed. Second, it is violently opposed. Third, it is accepted as being self-evident.
Arthur Schopenhauer, 1788–1860

OVERVIEW

Advances in medicine follow advances in biology. Most biomedical research improves our understanding of the details of known biologic processes. For example, once it was known that cancer could result from mutations in single genes in single cells, discoveries of mutated genes in cancer cells abounded and dozens have now been identified. Fundamentally new biologic discoveries occur rarely, however, and are dependent upon new ways of thinking and new technologies. A remarkable number of fundamentally important processes, such as embryonic development and memory, are poorly understood and await new insights. The lack of understanding of embryonic development is especially puzzling given its key importance to every living thing. Those processes that ensure reproduction of the organism are the most important to survival and, hence, may be the most robust and redundant.

Most biomedical research focuses on adult diseases, with interest in organ development principally as clues to disease causes. Adult diseases can be broadly divided into three problems at the cellular level: too much cell division, too little cell division, and defective cell function. Too much cell division results in tumors, which are sometimes cancerous. Too little cell division results in the inability to repair damaged tissues and organs, such as severed spinal cords and weak heart muscle. Defective cell function results in diseases such as hemophilia (lack of synthesis of blood clotting factors by liver cells) and diabetes mellitus (lack of insulin production by pancreatic cells). Historically, treatments for adult diseases have been surgery or pharmacologic intervention. For example, major research efforts have developed the methods for heart transplantation to "cure" weak hearts. Dozens of drugs have been developed to correct abnormal heart rhythms and support heart muscle weakened by high blood pressure or heart attack. The concept of repairing weak heart muscle cells by encouraging the growth of new muscle cells has, until recently, not been considered possible. This new way of thinking about treating disease has brought **stem cells** to the forefront. The term stem cells is used to describe those cells that serve as a normal reservoir for new cells needed to replace damaged or dying cells. As will be described in the following chapters, a fundamental characteristic of stem cells is the lasting ability to multiply when called upon. Stem cells are broadly divided into four groups: adult stem cells, fetal stem cells, embryonic stem cells, and nuclear transplant stem cells (Table 1.1).

Stem cell: a cell capable of unlimited cell cycles with the capacity to cease being a stem cell and become one of a wide variety of specialized cells that can no longer divide.

Adult Stem Cells

After their formation in the fetus, some tissues and organs continue to maintain a population of stem cells throughout childhood and into adulthood. Even though these cells appear early in development, they are called *adult stem cells*. One example is blood vessels that readily repair damage and can actually regenerate. Another example is spermatogenesis. The fetal testes is populated with stem cells that are not actually stimulated to produce more cells until puberty, but once sperm production is initiated, it continues throughout the life of the male. Billions of new sperm are produced daily. Another example is bone marrow, which maintains a population of blood cell stem cells that also produce billions of new blood cells each day. Cells that line the stomach and intestines are also regularly replaced. These examples illustrate fundamental features of adult stem cells: they maintain the ability to divide throughout life and give rise to specific cell types. Blood vessel stem cells give rise to blood vessels, but cannot give rise to sperm, and spermatogonia cannot give rise to blood. Hence, the developmental potential of adult stem cells is restricted.

Table 1.1

Classes of Stem Cells

Source	Description	Examples
Adult organs	Reservoir of cells partially committed to type of tissue in organ.	Liver, bones, bone marrow, lining of gut, and spermatozoa.
Fetal tissues	Precursor cells more abundant in growing organs; some uncommitted cells.	Heart muscle cells, brain cells, germ cell precursors to sperm and eggs.
Blastocysts	Inner cell mass cells of blastocysts created either by chemical/electrical stimulation of unfertilized eggs to produce **parthenotes**, or by infertility procedures with sperm and eggs.	Stem cells from parthenotes are under investigation; embryonic stem cells have the potential to give rise to all cell types.
Nuclear transplant blastocysts	Derived from eggs stimulated to divide following complete exchange of genetic information.	Under investigation; process related to animal cloning.

This restriction in developmental potential of adult stem cells is the essence of the need for new sources of stem cells. Those tissues and organs that do not maintain a population of stem cells throughout life cannot repair themselves. Why some tissues maintain stem cells and others do not is not understood, and is an area of active research. Tissues deficient in stem cells include brain, heart, spinal cord, eye, and kidney. There have been recent reports that a small population of stem cells also may be present in these tissues, and the challenge to scientists is to isolate them in the laboratory and encourage their multiplication into sufficient numbers to be therapeutically useful. One hope is to be able to accomplish this from a biopsy of the failing organ. Another hope is to be able to isolate stem cells from adult organs donated by individuals before their death for such a purpose. The enormous human and economic value of creating a bank of heart muscle stem cells or kidney stem cells, which contain enough genetic varieties to tissue match to all persons with failing hearts or kidneys, is clear and an active area of research.

That this may one day be possible is exemplified by bone marrow transplantation, which is, in fact, ongoing therapeutic use of adult stem cells. Dr. E. Donnell Thomas shared the Nobel Prize in Medicine in 1990 for his pioneering work in bone marrow transplantation, which began 50 years ago (see sidebar). Historically thought of as tissue transplantation, considerations of the role of bone marrow in blood formation reveal marrow transplantation is essentially bone marrow stem cell therapy.

Blood functions to carry oxygen, nutrients, hormones, and immune responsive cells to all parts of the body. Carrying out these functions requires continual cell death and replacement for several reasons. Bleeding or donating a unit of blood, for example, stimulates blood cell stem cells to multiply in the marrow to replenish the lost blood cells. Another example is replacement of the blood cells lost as a result of fighting an infection. The process of replacing the cells in peripheral blood actually involves several intermediate stages of **cell differentiation** following division of the stem cell (Figure 1.1). A fundamental difference between the bone marrow stem cell and the differentiating blood cells is the loss of the ability to divide again. There may be several stages of differentiation; some stages can divide again and others cannot. Cells not fully mature, but fully committed to one type of differentiated cell, are termed **precursor** or **progenitor cells** rather than stem cells, but the distinction between stem cells and precursor/progenitor cells is not always clear.

Cell differentiation: the process of becoming a fully mature, nondividing cell with specialized expression of gene products needed to carry out specific tissue functions.

As will be discussed in Chapter 12, bone marrow is no longer the only source of blood cell stem cells. Under certain conditions, venous blood can be enriched with bone marrow stem cells that can then be isolated by taking blood from a vein instead of by drilling a hole in a bone. This advance has not only improved our understanding of the equilibrium between bone marrow and peripheral blood, but has also markedly improved the availability of blood stem cells and made treatment far less painful.

Fetal Tissue Stem Cells

By approximately 3 months of pregnancy, organs are formed (described in Chapter 9). The developing human fetus is only a couple of inches long and weighs about 1 ounce, but primitive organs are in place. During the remaining 6 months of pregnancy, the organs enlarge and develop the ability to function independent of the placenta. Because the fetus is growing rapidly, all tissues and organs, including the brain, contain stem cells. It is for this reason that stem cell researchers are interested in studying fetal tissues. As described in Chapters 9, 12, and 13 much progress has been made, but many questions remain.

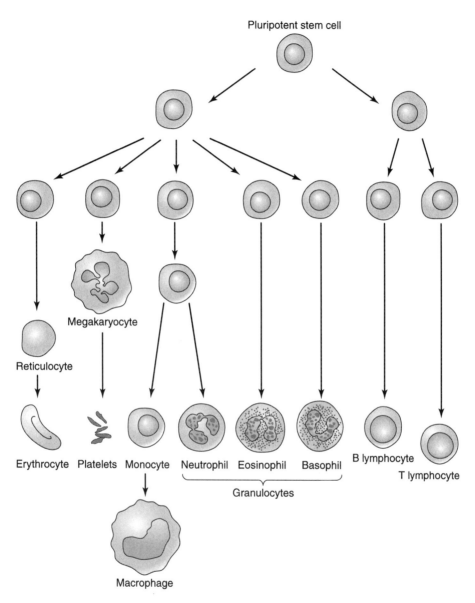

Figure 1.1   Blood cell pathways.

Fetal tissue research has long sparked debate between scientists and some social groups. The fear seems to be that some pregnancies will be terminated (and perhaps created) just for research. It is to the credit of our society that this fear exists. The concern speaks to the high value we place on individual human life, and our interest in

BONE MARROW TRANSPLANTATION

E. Donnell Thomas

In 1990, E. Donnell Thomas, M.D., shared the Nobel Prize in Physiology or Medicine with Joseph E. Murray, M.D. Both men pioneered medical therapies involving transplantation of tissues from one person to another. Dr. Murray pioneered kidney transplantation as a treatment for kidney failure, and Dr. Thomas pioneered transplantation of bone marrow cells as a treatment for leukemia and aplastic anemia. A fundamental difference between kidney transplantation and bone marrow transplantation is that the transplanted kidney is fully formed, but bone marrow cells must continue to proliferate and give rise to peripheral blood cells throughout the life of the transplant recipient. Thus, bone marrow transplantation, initially explored over 60 years ago, was the first form of stem cell therapy.

During Dr. Thomas' student years at Harvard Medical School (1943–1946), he became interested in leukemia and bone marrow. Research on the health effects of exposure to radiation (primarily X-rays) had revealed that death resulted from loss of cells in the bone marrow, which led to severe anemia and loss of immune function. One possibility that intrigued Dr. Thomas was that the irradiation destroyed "stimulating factors" that promoted growth of bone marrow cells, and he spent a year at the Massachusetts Institute of Technology studying factors that stimulated yeast cell division. In 1955, work by Main and Prehn demonstrated that a mouse treated after radiation exposure by transplanted bone marrow also did not reject skin grafted from the marrow donor. Thus, the immune system of the irradiated mouse had been reconstituted from the transferred bone marrow, providing strong evidence that bone marrow contained blood and immune system stem cells.

From 1955 to 1962, while an affiliate of Columbia University, Dr. Thomas established the basic concepts for successful bone marrow transplantation using dogs as a model system for humans. In 1963, he was recruited to the Department of Medicine at the University of Washington School of Medicine in Seattle, where he compiled a team of physician researchers who by the early 1970s had demonstrated that patients with leukemia, aplastic anemia, or genetic diseases could be cured by marrow transplantation. Such procedures have now become routine medical care, and thousands of successful and curative bone marrow transplantations are conducted annually worldwide.
(www.nobel.se/laureates/medicine)

providing social defenses for those members of our society who cannot defend themselves. In many respects, this fear represents substantial progress in our sense of humanity, and is the antithesis of the values of terrorists for whom individual life is less important than an ideologic goal.

The social fear also relates to distrust of science and technology. There is an underlying belief that appropriate oversight of fetal tissue research cannot be achieved because scientists are too zealous in their pursuit of their research interests to be mindful of restraints. This underlying belief is also fueling the embryonic stem cell debate. Scientists must accept responsibility for the distrust of some elements of society. The distrust stems from many areas, including developments that are clearly unsafe, like nuclear weapons, and seemingly unnatural, like genetically engineered foods and assisted reproduction technologies. Profit motives, as exemplified by the pursuit of patents, add to the sense of distrust of scientists. Scientific language is replete with abbreviations for complex cell processes that makes it difficult for nonscientists to follow the research, even though many of the concepts are easily understandable. The communication gap heightens the distrust. Bridging the communication gap between science and society will be essential for complete acceptance of fetal and embryonic stem cell research.

It is also clear, however, that some social groups seeking to advance their own specific views attempt to misrepresent the goals of scientists. For example, the concept that fetal tissue and embryonic stem cell research cannot be conducted ethically is often advanced by groups who also seek to outlaw abortion. Some of those groups conduct large, active lobbying programs in Congress to promote federal laws against all forms of embryonic stem cell research, claiming similarities to abortion. Fully informed decisions by ethicists, the general public, and legislators require understanding the biology of human embryonic development and stem cells.

Embryonic Stem Cells

Before organs are formed, embryos of all species are comprised of a collection of cells with potential to give rise to many different organs and tissues. Such cells with plural potentiality are termed **pluripotent** stem cells. First derived from mouse embryos, as described in Chapter 8, embryonic stem cells were found to have remarkable plasticity. They divide endlessly in laboratory culture dishes and maintain the ability to differentiate into numerous types of cells when exposed to the appropriate growth factors. For example, nerve growth factor, a protein hormone, can induce differentiation of stem cells in culture into mature nerves, **neurons**, capable of transmitting electrical and chemical signals remarkably similar to nerves in the body (Figure 1.2). This holds great promise for repair of neurologic diseases such as

Figure 1.2 Stem cells derived from the blastocyst of an activated monkey egg can differentiate into functioning neurons. (Modified from Cibelli, J. B., et al. [2002]. Parthenogenetic stem cells in nonhuman primates. *Science* 295: 819.)

Table 1.2				
Embryos Created by Assisted Reproductive Technologies in the U.S.				
Year	Cycles of Hormone Treatments	Estimated Fertilized Eggs/Cycle	Average Number Transferred	Estimated Number of "Leftover" Embryos
1997	71,826	5 to 10	3 to 4	143,652 to 502,782
1999	87,000	5 to 10	3 to 4	176,000 to 609,000

Values are estimates because the Centers for Disease Control does not report data relative to numbers of eggs harvested and fertilized.

Alzheimer's, Parkinson's disease, spinal cord injury, and retinal degeneration. The problem is finding a source of embryonic stem cells.

The proliferation of "leftover" human embryos as a result of aggressive hormonal treatment of women seeking to achieve pregnancy through assisted reproductive technologies (Table 1.2) has raised the possibility of deriving embryonic stem cells from human embryos by the same methods as applied to the mouse embryonic stem cells, which will be described in Chapter 11. This area of research has been met with social resistance similar to fetal tissue research, and has sparked additional controversy related to when, exactly, an embryo becomes human. The point at which human life is thought to begin varies among religions, so religious views play a substantial role in the debate.

From the standpoint of medical treatments, embryonic stem cells have the same tissue rejection concerns as stem cells derived from fetuses. Tissue compatibility problems may be less severe for fetal and embryonic stem cells than with adult organ transplants, but the cells will still express the proteins encoded by the genes of the embryo when they undergo differentiation. Fortunately, new advances in the manipulation of cells and eggs, described in the following chapters, holds the promise of creating stem cells for each person with a need without creating a unique embryo. One problem, however, is that some of the new methods are the same as those used for cloning individuals. Social groups already opposed to fetal tissue research, and/or the generation of embryonic stem cells from abandoned human embryos, are truly aghast at the concept of pursuing technology that could also clone a human. Informed decisions for or against, however, require a thorough understanding of the technology.

Nuclear Transplant Stem Cells

As will be described in the following chapters, a fundamentally important biologic advance was the realization, in the early 1900s, that all human cells contain all chromosomes. In this way, each cell of each tissue and organ is a replicate reservoir of all the genes of each person. The single exception to this paradigm are mature sperm, which contain half the number of chromosomes of all other cells, leading to equal populations of X-bearing (female) and Y-bearing (male) sperm.

This led to the question of whether or not the chromosomes from adult cells could support the development of a completely new individual, theoretically identical to the adult. If true, then the cell differentiation needed for organ formation would be the result of cellular controls on the expression of individual genes. If false, then the cell differentiation noted in organ development may result from irreversible changes to the genes. There was also the possibility that the genes of some types of cells, but not all cells, were irreversibly altered. For example, it seemed possible that cells from those tissues with regenerative capacity, such as bone marrow stem cells or the lining of the stomach and intestines, might not contain irreversibly altered genes, whereas spinal cord nerves that do not regenerate may contain irreversibly altered genetic information.

One way to test these possibilities, as described in Chapter 7, was to transfer a nucleus into an egg whose chromosomes had been removed (**enucleated**) to determine if the transferred nucleus could guide the development of an individual. Since this process involves manipulation of single cells, the egg and the cell donating the nucleus, it was termed **cloning**. The term stems from the ancient Greek *klon*, which means twig, and refers to the fact that a twig could give rise to an identical tree. Ultimately, as described in Chapter 7, it was demonstrated that adult nuclei retain the potential for guiding the development of an entire animal, although this potential has not been demonstrated for cells from all adult tissues. The advantages of cloned animals to breeders, researchers and pet owners are clear and even compelling, but the real power of the technology lies in the creation of stem cells for treatment of diseases.

Enucleated: the process of removing cellular genetic information either in the form of an interphase nucleus or as chromosomes.

Such nuclear transfer stem cells would be genetically identical to the adult and capable of giving rise to replacement cells for all body tissues. Gene replacement techniques could correct genetic defects in the stem cells to enable them to correct inherited disorders such as hemophilia and some forms of diabetes. The benefit to sufferers of now incurable diseases is clear.

The challenge before us is to provide the understanding among opposing social groups, religious interests and legislators to allow the biology to move forward with societal controls that will not deny these powerful treatment options to the millions suffering from incurable diseases.

Human Cloning

Concerns about cloning a human range from religious views that it is unholy, to fear of eugenics and the creation—in vast "clone farms"—of genetically selected individuals, generally referred to as some form of superbeing. The arguments and fears surrounding cloning serve to force society to examine what it truly values. The strong opposition at this time to even developing the technology speaks for the high value we place on human life and its attendant dignity. However, the opposition also raises significant inconsistencies in current social practices, such as the death penalty, and the tolerance of polygamy. Is it mere coincidence that Ossama Bin Laden, the terrorist who seeks to impose his way of life on the entire world, is one of 50 or so children from a polygamous father clearly seeking to enrich humanity with his own genetics? From the Old Testament's Abraham to modern day Josiah Smith, man has manifest a passion to pass on his genes. It needs to be recognized that to some individuals, cloning may represent simply a more advanced way to achieve this, not a new desire.

There are many reasons besides social and religious concerns not to clone a human. The experience with animals has shown that cloned individuals have a high incidence of failure to thrive. Since this is not true of identical twins, who are clones of each other, there is a problem with the current technology. Given this knowledge, it is at the least irresponsible to attempt to clone a human until the problems are solved in animal models. However, given the notoriety it will bring, it is likely to be attempted within the forseeable future. Society then will be forced to face a new set of fears. On the positive side, the technology could be modified to allow reproduction by men who have no sperm and women who have no eggs. This possibility could be viewed as providing a social and genetic balance against those highly fertile individuals clearly bent on proliferating their own gene pool.

Additional Readings

Green, R. A. (2001). *The Human Embryo Research Debates*. New York: Oxford University Press.

Kirschstein, R., and Skirboll, L. (2001). *Stem Cells: Scientific Progress and Future Research Directions*. Bethesda, MD: National Institute of Health. www.nih.gov

CHAPTER 2

The Dividing Cell

The uniformity of earth's life, more astonishing than its diversity, is accountable by the high probability that we derived, originally, from some single cell, fertilized in a bolt of lightning as the earth cooled.

Lewis Thomas, The Lives of a Cell

OVERVIEW

Robert Hooke coined the term *cell* from observations of cork made with a simple microscope in 1665. A few years later, Antony van Leeuwenhoek described sperm, red blood cells, and some bacteria. However, it was not until the 1830s that the microscopy studies of Theodor Schwann and Matthias Schleiden reached the same conclusion for animals and plants, respectively: all organisms are comprised of individual cells. Much has been learned in the ensuing 170 years, but much remains to be discovered.

Every advance in microscopy was accompanied by new discoveries about cells. Their most obvious features (Figure 2.1) are their outer (**plasma**) membrane, their **nucleus**, also enclosed with a (nuclear) membrane, and the fact that cells only arise from other cells. In this regard, all human life is a continuum. Further advances in microscopy also revealed other structures in the cell cytoplasm, including **mitochondria**, which generate energy, and **ribosomes**, which organize protein synthesis, as well as the **Golgi apparatus**, which is responsible for processing newly synthesized proteins and other large molecules synthesized by the cell. Importantly, each of these structures must also be replicated for division into the daughter cells.

Tissues and organs are made up of individual cells with specific functions arranged in a specific order. Skin, for example, is comprised of several thin layers of cells that are nourished by small blood vessels (Figure 2.2). Hair follicles, and

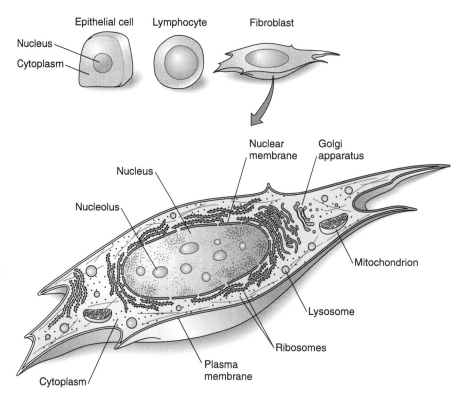

Figure 2.1 Prototype human cells.

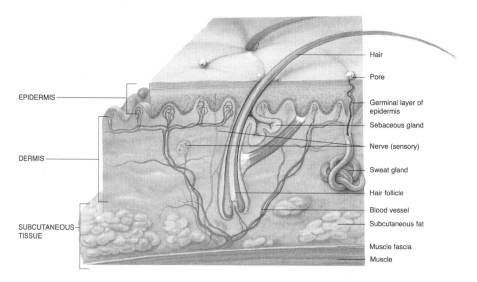

Figure 2.2 Layers of skin.

oil and sweat glands are other types of specialized cells that contribute to the function of skin. Collectively, the cells comprise the elastic body covering that is essential for life. Skin is an example of a tissue that contains a significant reservoir of stem cells and is a likely source of nuclei for nuclear transfer stem cells. Understanding stem cells, including those created by nuclear transplant technology, requires understanding fundamental properties of cells.

The Cell Cycle

Microscopic studies revealed two obvious stages in a cell's life: the stage when the nucleus is intact (**interphase**), and the stage when the cell is dividing and the nuclei are not visible (**mitosis**). For the most part, cell division brings about two cells (**daughter cells**) identical to the original (Figure 2.3). Thus, it has long been appreciated that a cell is capable of completely duplicating itself. This fundamental observation in and of itself implies inheritance of all genetic traits from one cell to the next, and lays the groundwork for genetic principles.

Gregor Mendel failed his exams to become a monk in the 1850s, but stayed at the monastery to conduct his famous white and purple pea-blossom experiments (Figure 2.4), which established the basic genetic concepts of **inheritance** and **dominance**. Genetic traits are inherited equally from both parents, and some genes that define those traits are more likely to be expressed (**dominant**) than other genes (**recessive**). For example, the purple blossom gene for color is dominant over the white blossom gene for color. These same principles of genetics have been shown to be true for mammals, including humans, and indicate that all genetic information inherited from each parent must somehow be present in sperm and eggs.

Inheritance: a gene passed on to offspring.

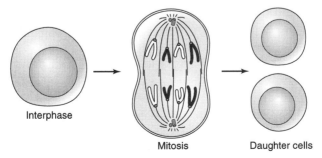

Interphase   Mitosis   Daughter cells

Figure 2.3   Cell division.

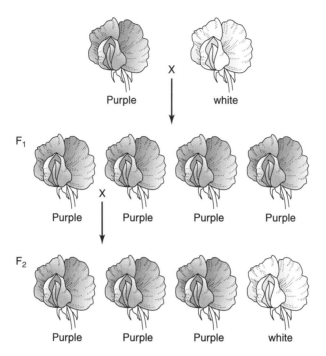

Figure 2.4 Gregor Mendel's pea blossom experiments.

Once it was known that all cells contain all genes, arranged on chromosomes, mitosis, therefore, must yield two genetically identical daughter cells with two copies of each chromosome. Increased resolving power of microscopes confirmed **chromosome pairing** and their precise division during mitosis (Figure 2.5). This suggested that the chromosomes contained within the nucleus might be genetic reservoirs. Interphase, the period of time during which a nucleus is visible, but chromosomes are not, is, therefore, the time during which the chromosomes are duplicated.

Chromosomes are comprised of the unique polymer, deoxyribonucleic acid (**DNA**), which encodes the genetic information. This discovery was made by Oswald Avery, Colin MacLeod, and Maclyn McCarty in 1944. Interestingly, the discovery was made in an attempt to understand the bacterium, *Pneumococcus*, that causes pneumonia. In the next decade, James Watson and Francis Crick published the structure of DNA as two highly organized polymers, ionically bonded together in a rope-like, helical structure (Figure 2.6). Their predicted structure was based on data obtained from X-ray crystallography studies by Rosalind Franklin and Maurice Wilkins.

The fundamentally important discovery that inherited traits are encoded by DNA, which is a polymer of specific sequences of four molecules joined together to form an enormous double-strand, led to an explosion of research on the mechanisms

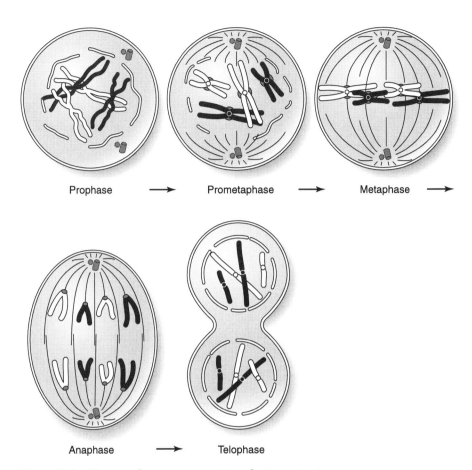

Prophase ⟶ Prometaphase ⟶ Metaphase ⟶

Anaphase ⟶ Telophase

Figure 2.5   Human chromosome pairing during mitosis.

of chromosome duplication (Figure 2.6). The work continues today because all of the details of replication of DNA are still not known. What is known is that duplication of all the DNA in each chromosome is necessary before the cells can divide. The period of DNA synthesis within interphase is termed **S phase** of the cell cycle.

"S" phase: the process of faithful duplication of each strand of DNA constituting each chromosome in a cell so as to yield two identical copies.

The advent of radioactive DNA precursor molecules allowed experiments to determine the duration of S phase, as well as provide clues about the process that leads to the faithful duplication of each DNA strand. It was discovered that there is a time interval, after cell division and the appearance of the new nucleus, before DNA synthesis begins. This interval is defined as **Gap 1 (G1)**. In addition, there is an interval of time between the end of S phase and the onset of chromosome formation. This interval

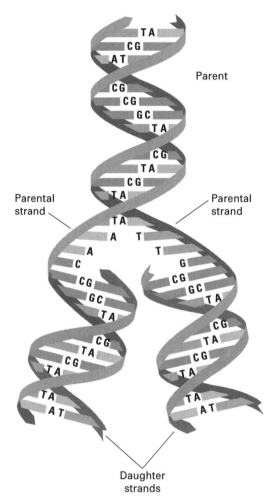

Figure 2.6   The double-stranded helical structure of DNA and replication of both of its strands.

is defined as **Gap 2 (G2)**. It was also discovered that each strand of the DNA molecule serves as the template for a new strand. Thus, it is important to note that each new chromosome has one old strand of DNA and one new strand of DNA (Figure 2.6). This process is called **semi-conservative replication** because one of the old strands is always maintained. The newly replicated double strands represent a duplication of the chromosome.

During Gap 2, the cell prepares for division termed **mitosis** or **M phase**. During early M phase, the DNA begins to rearrange itself according to individual chromosomes

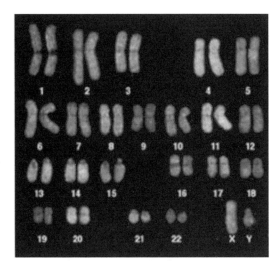

Figure 2.7   Human chromosomes. (Courtesy of Johannes Wienberg and Thomas Reid)

and discontinues its association with the scaffolding in the nucleus in preparation for mitosis. Each duplicated chromosome becomes compacted in a process termed **chromosome condensation** to describe the tightening of the DNA strands into coiled structures strong enough to withstand cell division without losing any genetic information. The duplicated chromosomes are termed chromatids because they stay joined at a central region, the **centromere**, which anchors the **mitotic spindle**, the strong protein fiber that pulls the chromatids apart at cell division.

Humans have 23 pairs of chromosomes—22 identical pairs termed **somatic chromosomes** and one pair of **sex chromosomes**, which is a pair of X chromosomes in the female, but an X and Y chromosome in the male, designated for their appearance (Figure 2.7). One chromosome of each pair comes from the mother, the other from the father. The terms somatic chromosomes and sex chromosomes are descriptive, not functional. No particular chromosome gives rise directly to any particular organ, and the genes for testis do not reside strictly on the Y chromosome. However, genes with related functions are sometimes found clustered together in chromosomes. For example, genes that encode the proteins displayed on the plasma membrane of cells that define "self" cluster on the long arm of Chromosome 6.

Somatic: all cells and tissues of an organism except sperm and eggs and their precursor cells, termed germ cells.

During mitosis, the chromatids are separated into the daughter cells, yielding new cells with two copies of each chromosome, one from the mother and one from the father (Figure 2.5).

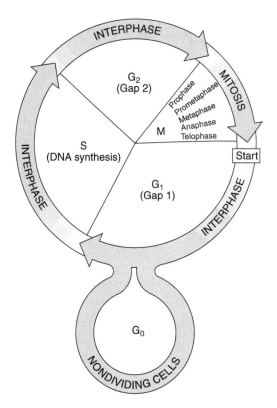

Figure 2.8 Schematic of the phases of a dividing cell.

A standard cell cycle is approximately 18 to 24 hours long, with a Gap 1 of 7 to 9 hours, an S phase of 6 to 8 hours, a Gap 2 of 2 to 4 hours, and a mitosis (M) phase of 1 to 2 hours (Figure 2.8). After cell division, some cells undergo additional changes in function to carry out specific tasks. This change in cell function is termed **differentiation**, and is usually accomplished in response to a specific factor that impacts the new daughter cell. For example, nerve growth factor is a protein that can bring about differentiation of stem cells into primitive neurons, as described in Chapter 1 (Figure 1.2). Cells that undergo differentiation usually do not divide again. Those cells actually exit the cell cycle and enter a mitotically quiescent phase termed Gap zero (**Gap 0**, or **G0**). Most cells that enter G0 never divide again, but some, like stem cells, can be recruited from G0 to reenter Gap 1. During this resting phase, the cell carries out its specific tasks as a differentiated cell in addition to housekeeping functions, such as respiration, and may grow a little, but it does not synthesize the proteins necessary to enter into a dividing cell cycle. The onset of cell division by most cells, including stem cells, requires a specific stimulus.

Stimulating Cell Division

The potato peeler accidentally removes a piece of finger skin. A cascade of emergency signals is immediately released by the damaged cells. A blood clot forms. Underlying skin cells (the dermis) knit together to seal the wound against fluid loss from below and infection from above. They turn on synthesis of keratins, the specialized proteins that toughen them to become surface cells (the epidermis). The stem cells beneath the dermis receive signals to begin to divide to replace the epidermis cells that are moving upward to become epidermis (refer to Figure 2.2). The process will take at least 24 hours because the skin stem cells have been resting in G0 and must awaken to enter G1. The stem cells will continue to produce daughter cells until the damaged area is repopulated.

A fundamentally important characteristic of stem cells is that only one, not both, of the daughter cells go on to differentiate; the other one must remain an undifferentiated stem cell to maintain the reservoir. Although there is still much to be learned, skin apparently has more stem cells than other body tissues, such as the heart, which is unable to replace damaged muscle cells within a few days. One possibility is that the heart does have stem cells, but their division is suppressed by other factors in the tissue that inhibit the cell cycle of all cells. Understanding all the factors involved in controlling cell division is an active area of research.

Molecular Biology of the Cell Cycle

Early work, which suggested that factors outside the nucleus, in the cell cytoplasm, controlled the cell cycle, came from studies of frog eggs, as described in Chapter 3. During the same period of time, three scientists—Lee Hartwell, Paul Nurse, and Tim Hunt—working with yeast and sea urchin eggs, respectively, discovered fundamentally important cytoplasmic controls on cell division. The three scientists shared the Nobel Prize in 2001 for their discoveries (see sidebar).

Progression through the cell cycle is now known to be regulated by a specific set of genes that encode proteins that can add phosphate groups to other proteins (Figure 2.9). Such proteins are **enzymes** called **kinases**. Enzymes are proteins that bring about chemical reactions at body temperature that would otherwise require heat. Phosphate groups have a strong ionic charge that either attracts or repels other groups and, in this way, changes the shape of proteins. Sometimes changing their shape causes enzymes to become more active, sometimes less active.

Kinase: a protein (enzyme) capable of bringing about the addition of a phosphate group ($PO4^=$) to an amino acid.

Two classes of proteins encoded by genes important to the cell cycle are **cyclins** and **cyclin-dependent kinases** (**Cdks**). All cells with nuclei, from yeast to human,

CELL CYCLE REGULATORS

Lee Hartwell, Paul Nurse, and Tim Hunt

In 2001, three scientists—Lee Hartwell, Paul Nurse, and Tim Hunt—shared the Nobel Prize in Physiology or Medicine "for their discoveries of key regulators of the cell cycle." In his Nobel Prize Presentation Speech, Professor Anders Zetterberg noted that "All living organisms on earth are descended from an ancestral cell that appeared about three billion years ago, and which has undergone an unbroken series of cell divisions since then. Each human being began life as one single cell—a cell that divided repeatedly to give rise to all one hundred thousand billion cells that we consist of. Every second, millions of cells divide in our body."

Dr. Hartwell did his undergraduate work at California Institute of Technology and his Ph.D. studies at the Massachusetts Institute of Technology. As a young researcher at the University of California, he began to use baker's yeast as his model organism. Although a microorganism, yeast has a nucleus that contains the yeast's chromosomes. The advantage of yeast as an experimental model is that it has a short cell cycle time. The short cell cycle time speeds up the formation of mutant organisms. By carefully characterizing mutant organisms, some of which were unable to divide at slightly higher than normal temperatures, Lee Hartwell described dozens of genes important to cell division. He termed these genes cell division cycle, CDC, genes. He noticed that the cell cycle only proceeded forward if the genes were expressed in a specific order. From this observation, he formulated the concept of cell cycle checkpoints. One gene, CDC28, proved to be essential for the initiation of each cell cycle.

Many checkpoints are in G1. For example, cells must grow to an appropriate size before initiating DNA synthesis. This requires appropriate nutrients and growth factors. Dr. Paul Nurse also used yeast for his cell cycle studies, but not baker's yeast. Dr. Nurse completed his undergraduate studies at the University of Birmingham and his graduate studies for his Ph.D. at the University of East Anglia in England. Dr. Nurse discovered that specific mutations in cell division cycle gene number two, Cdc2, led to cells that either did not divide or divided too early. He hypothesized that Cdc2 was a principal regulator in the cell cycle, and, in fact, controlled the cell cycle. By incorporating human genes into mutant yeast, he proved that the human gene analogous to the yeast gene (a homolog) could perform the correct function in yeast. Thus, Cdc2 in the yeast used by Dr. Nurse was the same gene as CDC28 in baker's yeast, and the same gene as human Cdc2. Remarkably, a gene responsible for initiating the cell cycle in baker's yeast had been conserved for more than one billion years of evolution, from one-celled organisms to man. This illustrates a fundamental biological principle: if it works, don't mess with it.

Dr. Tim Hunt completed both his undergraduate and graduate studies at the University of Cambridge in England. Dr. Hunt worked with sea urchin eggs as a model organism. In 1982, he discovered another gene that was expressed immediately before cell division, but not expressed at all immediately after cell division. He named the gene cyclin and discovered its existence in other types of cells as well. Soon thereafter numerous cyclin genes were discovered and the proteins they encode form complexes with the proteins discovered by Dr. Nurse and Dr. Hartwell. These fundamentally important discoveries paved the way for major advances in understanding cellular controls.

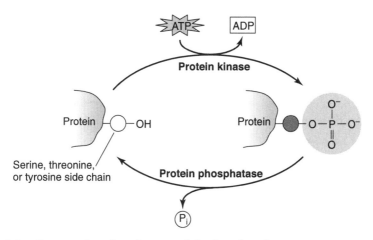

Figure 2.9 Protein phosphorylation and dephosphorylation.

contain genes that encode very similar protein families (Table 2.1). As the names imply, cyclin-dependent kinases are not active until they bind to their respective cyclin, a protein that is not an enzyme on its own. There are several related, but different, cyclins. For example, Cyclin E activates Cdk2, which is responsible for the transition of the cell from G1 to S phase. Since cyclin E is degraded immediately after it activates Cdk2, it must be synthesized by the cell during each G1 phase. The cellular controls that turn on and turn off synthesis of the proteins encoded by the cyclin genes are clearly fundamentally important to whether or not a cell is capable of cell division. This is an active area of investigation.

Similarly, cyclins A and B regulate the activity of another kinase, Cdk1, which is essential for the transition from S phase to G2. The observation that separate sets of cyclins and Cdks function at defined points in the cell cycle established the possibility of **checkpoints** that could stop the progression through the cell cycle if an error has occurred. For example, an important checkpoint occurs in G1 before the initiation of S phase. DNA damage is repaired before replication of the DNA begins. A protein termed p53 mediates the G1 checkpoint. Interestingly, many human cancers contain mutations in the p53 gene that lead to synthesis of a defective p53 protein that is not capable of mediating

Table 2.1

Mammalian Cell Cycle Regulators

Cell Cycle Phase	Cyclin	Cdk
G1	A, D1, D2, D3, E	Cdk2 (Cdc2)
	D1, D2, D3	Cdk4
	D2, D2, D3	Cdk5
S	A, B	Cdk1
G2 to M	A, B	Cdk1

the checkpoint. This, therefore, allows cells with DNA damage to initiate S phase. In this way, defective cells characteristic of cancerous tumors are allowed to accumulate. Such cells have escaped the natural checkpoints for cell division, and also usually fail to carry out the normal functions of the tissue in which they arose.

Another important checkpoint is at the end of S phase. DNA replication requires an enormous complex of enzymes, **DNA polymerases**, and other accessory proteins. They are synthesized during G1 and destroyed immediately after S phase. In this way, the cell maintains strict control of the process of DNA synthesis. As will be seen in later chapters, an exception to this rigid control exists in eggs and early embryos, which exhibit unusual cell cycles, the reasons for which remain a mystery. It is obviously critically important to the cell line that mitosis not begin until DNA replication is completed. Moreover, should something happen to the cell during S phase, such as exposure to X-rays or ultraviolet light, damage to the DNA would need to be repaired before S phase could be completed. In circumstances where the damage cannot be repaired, cell death frequently occurs as a result of stimulation of specific cell death pathway enzymes.

Progression from G2 to M phase requires activation of Cdk2 (sometimes denoted Cdc2) by binding to cyclin B, which leads to an activity originally identified as **maturation promoting factor (MPF)** in frog eggs, described in Chapter 3. Cyclin B synthesis begins during S phase. As it is synthesized, it forms a complex with Cdk2. The complex is immediately phosphorylated at three amino acids on Cdc2 (threonine 14, tyrosine 15, and threonine 161) to maintain its inactive state (Figure 2.10). As discussed previously, adding phosphate groups (phosphorylation) can either activate or inactivate enzymes. Enzymes that remove phosphate groups are termed **phosphatases** (Figure 2.9). Cdc25 is the phosphatase that brings about the activation of the cyclin B/Cdc2 complex (MPF) by removing the phosphate groups from threonine 14 and tyrosine 15. The active complex is a kinase that phosphorylates several proteins essential to M phase, including other kinases and structural proteins, such as **Histone 1**, the protein that complexes with DNA and whose rearrangement is necessary for the formation of chromosomes, the **nuclear lamins**, which lead to dissolution of the nuclear membrane, and **microtubules**, which lead to the formation of the mitotic spindle.

Phosphatase: a protein (enzyme) capable of removing a phosphate group ($PO_4^=$) from an amino acid.

A checkpoint also exists during mitosis to ensure equal division of the chromosomes into each daughter cell. Once the chromosomes are lined up on the mitotic spindle, MPF must be inactivated to allow mitosis to proceed. This is accomplished by degradation of cyclin B by a pathway specifically activated by MPF at the initiation of M phase. Thus, MPF ensures its own finite lifetime specifically to orchestrate metaphase. The decline in MPF activity allows the formation of two new nuclei and a division of the cell cytoplasm and plasma membrane.

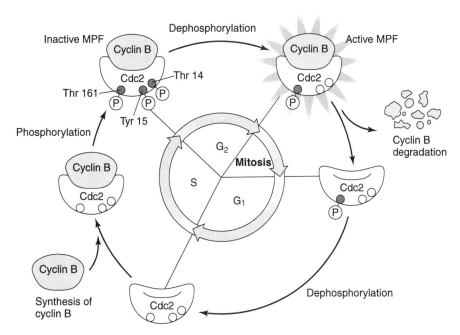

Figure 2.10 Regulation of maturation promoting factor.

In summary, duplication of a cell into two daughter cells is under the control of highly conserved genes. Their function is to optimize the possibility that two daughter cells, identical to the parent cell, will be formed. The process involves growth of the cell in size, duplication of the cellular organelles, synthesis of the large complex of DNA polymerases and accessory proteins responsible for faithful replication of the DNA, DNA synthesis, the formation of chromosomes and their accurate division into the two halves, and cell division, followed by the formation of new nuclei. In no cell are these processes more critical than in the stem cell, which serves as a life-long reservoir of cells capable of self renewal.

Additional Readings

Elledge, S. J. (1996). Cell cycle checkpoints: Preventing an identity crisis. *Science* 274: 1664–1672.

Oostendorp, R., Audet, T., and Eaves, C. (2000). High-resolution tracking of cell division suggests similar cell cycle kinetics of hematopoietic stem cells stimulated in vitro and in vivo. *Blood* 95: 855–862.

Weinert, T. (1998). DNA damage and checkpoint pathways: Molecular anatomy and interactions with repair. *Cell* 94: 555–558.

PART II

Egg Specific Functions

CHAPTER 3

The Egg

When you start with a portrait and search for a pure form, a clear volume, through successive eliminations, you arrive inevitably at the egg.

Pablo Picasso

OVERVIEW

The egg is wondrous. One of its most noteworthy features is size. The fully grown human egg has a diameter of 110 to 120 microns, yielding a volume of approximately 900,000 cubic microns. In comparison, a human white blood cell has a diameter of about 20 microns, yielding a volume of approximately 4,000 cubic microns. The volume of a sperm head, in contrast, is approximately 250 cubic microns. Thus, an egg is nearly 250 times larger than a white blood cell, and nearly 4,000 times larger than a sperm head.

The nucleus of the egg is also huge. One of the first structures noted by early microscopists, it was termed the **germinal vesicle** (Figure 3.1), a term

| **Germinal vesicle:** the large nucleus of an oocyte.

still in common use today. It is uniquely characterized by a large, prominent **nucleolus**, which is the organizing region for ribosomal RNAs. Together, the germinal vesicle with its large nucleolus resembles a giant belly button in the center of the egg. With a diameter on the order of 50 microns, the volume of the germinal vesicle is approximately 65,000 cubic microns, more than ten times the volume of a white blood cell. This huge nucleus provides ample open scaffolding for ease of expression of egg (**oocyte**) genes.

The egg also has an unusual and unique cell cycle. Given the importance of reproduction to the species, it is reasonable to assume there are multiple

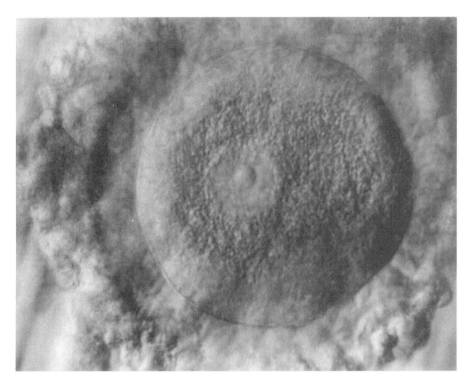

Figure 3.1   Germinal vesicle stage human egg.

safeguards to ensure an adequate supply of eggs with the potential to give rise to new offspring. Many aspects of egg development remain mysterious, however, principally because unlike other cells, laboratory methods have not been developed to support the life cycle of eggs in culture. This fundamentally important roadblock is the largest hurdle faced by nuclear transplant stem cell scientists. In the absence of a laboratory culture system to generate oocytes, the sole source of human eggs at the time of this writing is surgical recovery from the ovaries of normal, healthy adult women.

Egg Growth

The supply of eggs in the ovaries of each woman is limited to the population established early in her fetal development. The development of the ovary is less well studied in humans than in mice; much of the information about human ovarian development is, therefore, extrapolated from mouse and from some monkey models. Although the timing is not known with certainty for humans, as described in Chapter 9, at some point before organs are fully formed, the precursors to what will become either eggs or sperm (collectively termed **germ cells**) gather together in an area termed the **gonadal ridge**. The germ cell precursors are, in fact, stem cells for sperm and eggs. The difference between stem cells for sperm and stem cells for eggs is that egg stem cells contain no Y chromosome. As described in Chapter 2, humans have 23 pairs of chromosomes in each cell—22 somatic chromosomes and one pair of sex chromosomes. In the female, the sex chromosomes are a pair of X chromosomes (for their symmetric shape), and in the male the sex chromosome pair has one X and one Y chromosome (see Figure 2.7).

> **Germ cells:** general term for sperm and eggs and their precursor cells.

The presence of a Y chromosome leads to sufficient production of testosterone by the developing male fetus to cause the gonadal ridge to develop into a testis and support the development of other male reproductive tract organs, such as the epididymis and seminal vesicles. In the absence of a Y chromosome, an ovary forms from the gonadal ridge along with other female reproductive tract tissues, such as **oviduct (fallopian tubes)** and uterus, described in Chapter 9. Egg stem cells, **oogonia**, divide continuously in the fetal ovary during early development. Oogonia are not significantly larger than somatic cells. For reasons not fully understood, oogonia stop dividing in the fetal ovary during the second trimester of fetal development and become **primary oocytes**. Primary oocytes are arrested in late G2 of the cell cycle. This is a highly unusual cell cycle arrest point and the reproductive strategy behind this arrest is not entirely certain. In this state, they have twice the normal amount of DNA (**tetraploid**), and are said to be in the **prophase** of meiosis. This is the stage at which **chromosome crossover** can occur, a process most thoroughly studied in yeast.

Each cell in the developing fetus has one chromosome from the father and one chromosome from the mother. This is the genetic paradigm first predicted by Mendel's pea blossom experiments, described in Chapter 2. Chromosome crossover allows balanced exchange of DNA sequences (and therefore genes) between two like chromosomes. Although they are not, strictly speaking, chromosomes during prophase, the DNA strands that comprise each chromosome somehow recognize each other and matching sections of the DNA line up in a close orientation. They are positioned such that the DNA strands are spliced open and reunited with a strand from the paired chromosome (Figure 3.2). Since there are four copies of each chromosome, there are multiple opportunities for chromosome crossover in the primary oocyte. This process leads to new chromosomes that contain genes from both the father and the mother. Since the process of chromosomal crossover is thought to be relatively random, each egg that completes prophase of meiosis could theoretically have a unique set of chromosomes.

The primary oocytes will not resume growth and maturation for at least a decade, at the time the girl enters puberty and a regular monthly pattern of egg maturation and

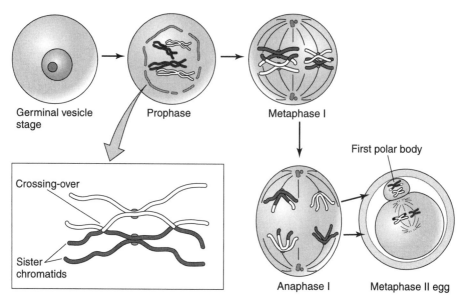

Figure 3.2   Chromosome crossing over.

release begins. It is not known with certainty if chromosome crossover is continuous in human oocytes throughout the long prophase of meiosis. It is known, however, that as soon as a population of oocytes is established, that population begins a decline in numbers that eventually culminates in menopause. Infant girls are born with on the order of one million primary oocytes. By the time she is approximately 50 years of age, the oocytes will all be gone. The best estimates are that after birth, oocyte death is essentially linear. This means that on the order of 20,000 eggs die each year, including the dozen or so which are ovulated.

Whether or not the ovulated eggs are selected from the remaining healthy cohort, rather than from those that have entered the death pathway, is not known with certainty. The fact that many more eggs can be fertilized than can actually give rise to offspring (Table 1.2) suggests that the selection process does not strictly exclude defective oocytes. There is ample evidence that following fertilization, most eggs undergo early cleavage, but fail somewhere during embryonic or fetal development. This suggests that many more eggs may have the potential to give rise to stem cells following nuclear transfer (**ovasomes**) than have the potential to give rise to babies. The human and medical value of being able to distinguish between cleaving eggs with the full potential to give rise to offspring and those with limited potential cannot be overstated. Research in this area has been essentially eliminated by the ruling of the U.S. Congress that "... such research is valuable, but will not be funded by taxpayer dollars."

Oocyte Meiosis

As described previously, oogonia stop dividing in the fetal ovary. This is in contrast to spermatogonia, which continue to divide throughout the life of the adult male. The oogonia undergo one final round of DNA synthesis, S phase, enter G2, grow larger, and enter the prophase of meiosis. As primary oocytes they are completely surrounded by a layer of **granulosa cells** (termed *nurse cells*), specialized hormone-producing cells in the ovary that play a major supportive role in oocyte growth, maturation, and eventual release from the ovary at **ovulation**.

Ovulation: the process of release of an egg from the ovary.

Maturation of a primary oocyte in preparation for ovulation is intimately tied to the monthly menstrual cycle. It is the result of wonderfully orchestrated communication between the ovary and two glands in the brain, the **hypothalamus** and the **pituitary** (Figure 3.3). The hypothalamus produces protein hormones whose target is the pituitary. The pituitary gland in turn synthesizes and releases more

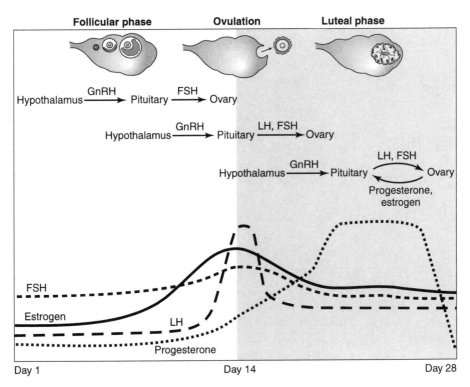

Figure 3.3 Hormonal communication between the brain and the ovary.

protein hormones. For example, the hypothalamus synthesizes and releases **growth hormone-releasing hormone**, which stimulates the pituitary gland to produce **growth hormone** to stimulate growth and development in children, including the long bones of the legs and arms. It seems as if once growth is essentially accomplished, the hypothalamus and pituitary turn their attention to the ovaries and the testes, collectively called **gonads**. The hypothalamus synthesizes and releases **gonadotropin-releasing hormone (GnRH)**, which stimulates the pituitary to release hormones to the bloodstream that stimulate the testis in boys and the ovaries in girls. The pituitary hormones are thus termed **gonadotropins**. They are released from the pituitary into the bloodstream in well-defined pulses throughout the day and night. The pulsatile hormone release persists until menopause in women and forever in men.

One of the gonadotropins is **follicle-stimulating hormone (FSH)**, which stimulates sperm production in the testis and growth and meiosis in primary oocytes in the ovary (Figure 3.3). It is not certain how eggs are selected for ovulation each month. It is theorized that several may be selected, but that one responds more vigorously than the others and eventually becomes dominant. FSH stimulates the granulosa cells surrounding the primary oocyte to divide and secrete estrogen. Estrogen in turn stimulates increased expression of receptors for FSH on the surface of the granulosa cells, thus capturing more FSH each time it is released into the bloodstream by the pituitary. In response, the oocyte resumes growth and expression of oocyte-specific proteins, including **zona pellucida** proteins, which form a protective coat around the giant cell (Figure 3.1). The zona pellucida plays several roles, as will be described in Chapter 6.

The cyclic recruitment of a primary oocyte to undergo maturation for ovulation that month begins within a day or two of the onset of menstruation, which signals that no pregnancy has occurred. In the absence of pregnancy, the granulosa cells in the ovary essentially cease synthesis of the steroid hormones, estrogen and progesterone. The decrease in steroid hormones stimulates the hypothalamus and pituitary to resume hormone synthesis. The cycle thus begins anew in response to FSH release by the pituitary. Approximately 10 days following the onset of menstruation, FSH release has stimulated sufficient estrogen production by the granulosa cells surrounding the primary oocyte to in turn stimulate the development of a fluid-filled, estrogen-rich sac termed an **ovarian follicle**. The advent of the ability to computer-render echoes from low-energy ultrasound waves has given rise to equipment that can display a computer-generated picture of the developing follicle without apparently damaging the egg. A probe that generates low-energy ultrasound waves from its tip, and senses their echoes, is placed either on the abdomen or in the vagina, directed toward the ovary, and the growing follicle is visualized (Figure 3.4).

Within the hormone- and nutrient-rich environment of the follicle, the oocyte cytoplasm stockpiles components it will need to resume development after being fertilized

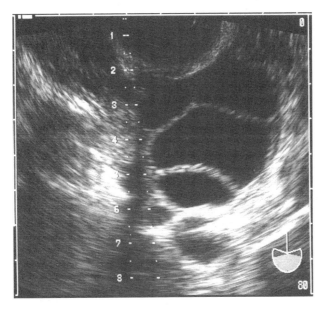

Figure 3.4   Growing ovarian follicles visualized by ultrasound. The dashed lines indicate measurement in centimeters.

by sperm. The stockpiling effort is assisted by projections of the granulosa cells, which extend through the zona pellucida into the egg cytoplasm. Numerous studies have shown that the granulosa cells thus serve as conduits for small molecules, such as amino acids and nucleic acids, which are shuttled directly from the follicular fluid into the egg cytoplasm. The oocyte is also stockpiling proteins and messenger RNAs as it completes its growth. An unusual feature of eggs that is not understood is the inactive state of oocyte mitochondria. Although they are abundant within the egg cytoplasm, they appear not to be actively involved in metabolism, a feature that distinguishes them from those in dividing cells. The role of maternal mitochondria in oocyte metabolism and potential for development is an area in urgent need of additional research.

Estrogen produced by the granulosa cells not only acts locally to stimulate the egg and other cells in the growing follicle, but also enters the bloodstream and stimulates the pituitary to decrease production of FSH and stimulate release of another gonadotropin, **leutinizing hormone (LH)**. By approximately day 13 of the menstrual cycle, LH pulses reach the same height as FSH pulses, which begin to decline (Figure 3.3). This circumstance initiates a dramatic cascade of responses in the follicle that ultimately leads to release of an egg poised for activation by sperm.

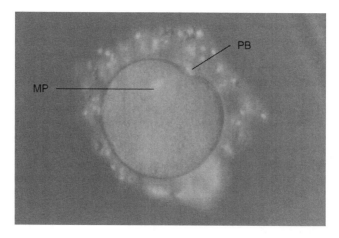

Figure 3.5   Egg arrested at metaphase II, surrounded by cumulus cells. PB = polar body; MP = metaphase plate.

LH transforms the granulosa cells from producing estrogen to producing progesterone and they stop dividing. Perhaps in response to the rise in progesterone, which is known to stimulate meiosis in frog eggs, the oocyte rapidly resumes meiosis. Protein and nucleic acid synthesis ceases. The huge germinal vesicle migrates toward the edge of the oocyte reminiscent of a primordial belly dance. Chromosome condensation begins and the germinal vesicle membrane dissolves in preparation for the formation of the **first metaphase plate**, the lineup of chromosomes attached to spindle fibers that, in the egg, appears adjacent to the plasma membrane, rather than in the center of the cell as in somatic cells. Almost as soon as the metaphase plate is formed, a unique, unequal cell division occurs, which results in the production of a small cell, the **polar body**, approximately the size of a somatic cell, which contains a complete set of chromosomes (Figure 3.5). In contrast to mitosis (see Figure 2.5), whole chromosomes, rather than individual chromatids, are segregated. This is termed **meiosis I**. Because of the crossing over between chromosomes that occurred in the prophase of meiosis, the egg and the polar body have both maternal and paternal genes rearranged on new chromosomes.

> **Metaphase plate:** the newly duplicated chromosomes separated opposite from each other by protein fibers.

Following the unequal cell division, in contrast to all other cell cycles, the nuclear membrane does not reform around the remaining chromosomes. They immediately undergo a rearrangement, which results in a **second metaphase plate** that contains the normal number of chromosomes of a somatic cell. Thus, separation of the chromatids at meiosis II yields two sets of single chromosomes, rather than chromosome pairs.

During this time, the granulosa cells immediately surrounding the egg retract their projections back through the zona pellucida, and synthesize and secrete hyaluronic acid, a viscous protein through which sperm must propel themselves. Their new appearance around the egg is less dense and termed the **corona radiata**. Because these granulosa cells have remained attached to the egg itself, instead of lining the walls of the follicle, they are termed **cumulus cells** (Figure 3.5). The follicular fluid has changed from estrogen rich to progesterone rich, and progesterone levels rise within peripheral blood. Ovulation is imminent. Within 1 to 2 days of the peak of LH, the wall of the follicle ruptures, releasing the egg-cumulus complex, approximately 1 centimeter in diameter. The complex is swept into the fallopian tube by pulsating cilia, developed within 2 or 3 days of ovulation to direct the egg toward the uterus.

Despite this elaborate, highly orchestrated preparation, the huge cell arrested at metaphase II is very fragile. If not activated within 1 to 2 days, the egg will pass unnoticed into the uterus and perish. As described previously, even if it is fertilized by a sperm, approximately three out of four eggs do not have the capacity to give rise to offspring. Those who believe that life begins at fertilization must accept the fact that 75% of such newly formed "lives" naturally perish. Understanding what leads to developmental success or failure is clearly an urgently needed area of research.

Assisted Reproductive Technologies

By the mid 1980s, a variety of medical treatments for infertility had been developed, discussed further in Chapter 4. One major advance was the development of microsurgical techniques to bypass and repair damaged fallopian tubes that blocked the passage of the egg from the ovary into the uterus. Another major advance was the discovery that the urine of menopausal women contains relatively high concentrations of gonadotropins. When all the eggs in the ovary have died, the cyclic production of estrogen and progesterone ceases and there is no feedback to the hypothalamus and pituitary to block the release of FSH and LH. As a consequence, menopausal women have high circulating levels of FSH and LH for a few years, some of which pass through the kidney with urine. An Italian company developed the laboratory methods for purifying gonadotropins from urine collected in convents by nuns. The availability of pharmacologic doses of such gonadotropins provided the opportunity to treat women whose infertility resulted from their own hormone imbalances. It also provided the opportunity to increase the levels of gonadotropins in hormonally normal women to stimulate more than one egg to resume maturation each month, thus paving the way to increase the number of eggs available for fertilization in laboratories.

Such **in vitro fertilization** (IVF) procedures have become the standard of care for a variety of infertility conditions, including low sperm counts. Additional gonadotropins are administered to the woman for several days early in the menstrual cycle to stimulate growth of multiple ovarian follicles. This clinical treatment is termed **controlled ovarian hyperstimulation** (COH). Importantly, the eggs recruited do not comprise a synchronous cohort and the growing follicles have a broad range of sizes (Figure 3.4). Nearly all eggs that have responded to the exogenous gonadotropins will, however, resume meiosis in response to a single injection of a gonadotropin that mimics the natural LH surge (Figure 3.3). In this regard, it is important to note that the capacity to resume meiosis II may precede full maturation of oocyte cytoplasm.

IVF: in vitro fertilization, fertilization of an egg by a sperm in a laboratory tissue culture dish.

Approximately 34 hours after administration of the LH surge, eggs can be collected directly from the ovarian follicles with the aid of ultrasound visualization. This timing is close to the LH-induced rupture of the follicle to allow meiosis to proceed within the follicle as in normal ovulatory cycles. Numerous clinical studies have revealed that more than 80% of eggs collected from women undergoing COH have reached metaphase II, will undergo successful fertilization by sperm, and initiate cleavage divisions (described in Chapter 6), but fewer than 10% of the fertilized eggs have the capacity to develop to a fetus. Unfortunately, at this time, there are no definitive indicators to identify the cleaving eggs that have the potential to develop into a fetus. As a consequence, most U.S. infertility clinics allow two or more early cleaving eggs to be returned to the uterus of the infertile woman for gestation on the hope that at least one will continue to develop. Although this practice has successfully improved the pregnancy rate per egg collection, it has also resulted in an epidemic of twins and triplets that are high-risk pregnancies. Given that more than 90,000 cycles of IVF are performed in the U.S. alone each year, with more than 30,000 babies born, this is clearly a major women's and children's health issue in urgent need of research.

Molecular Biology of Oocyte Meiosis

As discussed in Chapter 2, early work suggesting that factors outside the nucleus, in the cell cytoplasm, controlled the cell cycle came from studies of frog eggs. Landmark publications in 1971 by Masui and Markert, and Smith and Ecker revealed that transfer of cytoplasm from metaphase II frog eggs into germinal vesicle stage frog eggs initiated meiosis in the germinal vesicle stage eggs. They coined the term **maturation promoting factor** (MPF), which was eventually purified by Maller in 1988. Molecular characterization of frog egg MPF revealed it was composed of two cell cycle proteins, Cdc2 and cyclin B, described in Chapter 2.

MPF: maturation (or mitosis) promoting factor, a dimer of Cdc2 (Cdk2) and cyclin B.

Studies that followed demonstrated the same pathways in mammalian eggs. Thus, cell cycle regulators are also factors important to the resumption of meiosis.

But unlike dividing cells, mammalian eggs, including human, undergo another arrest at metaphase II. In the same series of experiments that led to the discovery of MPF, Masui and Markert transferred cytoplasm from a metaphase II-arrested frog egg into cleaving frog embryos and observed an arrest of embryo cleavage at M phase of the cell cycle. This indicated that metaphase II arrest was also controlled by cytoplasmic factors, which the authors termed **cytostatic factor (CSF)**.

Subsequent work in several laboratories (see sidebar) led to the identification of a kinase, cMos, as an important component of CSF. cMos was originally identified as a cancer-causing gene (vMos) in a virus (Moloney sarcoma virus) that leads to sarcomas in mice. The normal cellular counterpart, cMos, was found to be uniquely expressed in ovaries and testis. Its physiologic role in the testis remains to be elucidated, but its role in the ovary appears confined to the egg.

As described in Chapter 2, MPF is a complex of cyclin B and Cdc2. Cyclin B is synthesized during S phase and complexes with Cdc2. Activation of the complex by Cdc25 leads to kinase activity with a broad spectrum of targets, including other enzymes and structural proteins. Importantly, one of MPF's targets for phosphorylation early in M phase is the degradation machinery for cyclin B, which normally occurs after the metaphase plate is fully formed. Thus, to arrest the egg at metaphase II, it is necessary to block the degradation of cyclin B. This is the function of cMos. It phosphorylates a kinase, **mitogen-activated protein (MAP kinase)**, which plays a central role in maintaining chromosomes in metaphase after the extrusion of the first polar body. MPF activity, as measured by the phosphorylation of Histone type I (described in Chapter 2), decreases slightly at the time the first set of chromosomes is extruded, then increases again to maintain the Metaphase II chromosomes, for up to a few days in human eggs. Therefore, a fundamentally important aspect of egg activation is the elimination of cMos activity. Although not yet identified, the factors that eliminate cMos activity and allow the destruction of cyclin B to occur must be stockpiled within the egg's cytoplasm, awaiting activation.

These profound cytoplasmic controls on the egg cell cycle serve to emphasize the importance of cytoplasmic stockpiles to the success of egg activation. Moreover, the egg-specific controls need to be entirely replaced by dividing cell cycle controls within a brief interval following egg activation to allow cell cleavages to proceed normally. As demonstrated by the original studies with frog egg cytoplasm, factors in metaphase II-arrested eggs can bring about the arrest of cleaving blastomeres. Studies in animal models, principally the mouse, have revealed a rapid degradation of the stockpiles of messenger RNAs in the egg beginning at the time of fertilization. Such degradation may be as essential to successful egg activation as new RNA synthesis from the cleaving blastomeres. This area is in urgent need of additional research.

cMOS ACTIVITY

Metaphase II Arrest

Two groups of cancer scientists, led by Debra Wolgemuth and George vandeWoude, respectively, reported an unusual finding in 1985. A cancer-causing gene, vMos, first identified in Moloney sarcoma virus, was expressed specifically in testis and ovaries. vMos was known to encode a protein kinase that led to death in the cells infected by the virus. Subsequent independent studies by two other investigative teams, led by Geoffrey Cooper and John Eppig, respectively, revealed that cMos was expressed by eggs and not granulosa cells, and that it played an important role in maintaining oocyte arrest at metaphase II. Results from the Cooper group indicated that inhibition of cMos protein synthesis actually resulted in oocyte activation, as demonstrated by pronuclear formation and cell cleavage; the results from the Eppig laboratory indicated that inhibition of cMos resulted in the loss of metaphase chromosomes, but did not lead to egg cleavage. The presence of active cMos was confirmed in human eggs by the Cooper group.

These observations suggested an important role for cMos in reproduction. To understand the role, two other teams of investigators generated mice engineered without the gene for cMos. Eggs collected from the cMos "knock-out" mice demonstrated the same ability to undergo spontaneous activation as originally observed by the Cooper team. Taken together, the data suggest either that the experiments in the Eppig lab failed to completely inhibit cMos synthesis, suggesting two possible roles for cMos, one maintaining metaphase arrest and a second blocking mitosis, or that successful egg cleavage once cMos is inactivated requires specific laboratory conditions. Importantly, however, the females with the cMos gene deleted were not sterile, although they were subfertile. Thus, cMos expression enhances reproductive success, but is not absolutely essential. Given the importance of reproduction to species survival, the existence of multiple families of genes that enhance reproductive success seems highly likely.

In keeping with the predictions, the ovaries of the females with cMos "knocked out" routinely contain dermoid cysts resulting from spontaneous resumption of meiosis within the follicle.

Additional Readings

Gilbert, S. F. (1999). *Developmental Biology*, 5th ed., Sunderland, MA: Sinauer Associates.

Hunt, P. A. (1998). The control of mammalian female meiosis: Factors that influence chromosome segregation. *J Assist Reprod Gen* 15: 246–252.

Sagata, N. (1996). Meiotic metaphase arrest in animal oocytes: Its mechanisms and biological significance. *Trends Cell Biol* 6: 22–28.

Wassarmann, P. M. (1987). The biology and chemistry of fertilization. *Science* 235: 553–560.

CHAPTER 4

The Activated Egg

Do you see this egg? With this you can topple every theological theory, every church or temple in the world.

Denis Diderot, 1769

The egg arrested in metaphase II and surrounded by thousands of cumulus cells is released at rupture of the ovarian follicle and swept into the fallopian tube like a giant bride with a million attendants. Stockpiled in the egg's cytoplasm are stores of messenger RNAs and proteins that must be released and activated in the precise cascade demanded by the oocyte cell cycle.

As discussed in Chapter 2, dividing somatic cells systematically synthesize and then degrade the components of the cell cycle according to each transition phase. For example, the many proteins necessary to replicate all the DNA in the chromosomes are synthesized at the G1 to S transition, and then most are degraded in G2 immediately following the checkpoint for DNA synthesis errors. Dismantling the DNA replication machinery ensures that DNA synthesis does not occur at other times in the cell cycle, thus maintaining accurate gene copy numbers. Similar strategies are in place for other cellular components whose actions are restricted to one phase of the cell cycle, such as the cyclins.

This is not the strategy adopted by the egg. Even though it is arrested at metaphase II, relatively high levels of DNA polymerase activity are detectable in mouse and rabbit eggs. There must be an evolutionary or reproductive advantage to having all components possibly needed for egg activation and early cleavages stored within the egg cytoplasm. The task for the egg, therefore, is to deploy its stockpiles in the systematic and precise manner needed for faithful DNA replication and cell division. Moreover, it is important to remember that eggs have two unique cell cycle arrests: one in M phase when the egg is at the germinal vesicle stage and arrested in the prophase of meiosis for up to four or five decades, and the other at the time of ovulation when the egg is arrested at metaphase II (discussed in Chapter 3).

The cell cycle arrest of the germinal vesicle stage egg is highly unusual because it involves a lengthy nondividing state usually reserved for G0 (discussed in Chapter 2), but the giant cell is actually arrested in early M phase. Germinal vesicle stage eggs from several species, including cow, sheep, mouse, and human, collected directly from growing ovarian follicles and placed in culture in the laboratory can spontaneously undergo migration and breakdown of the germinal vesicle and resume meiosis. It is important to note that although most eggs that spontaneously resume meiosis in culture can undergo activation, including fertilization, they frequently fail following one or two cleavage divisions. This indicates that the capacity to resume meiosis precedes full maturation of the egg. The capacity to support fertilization follows, but since many fertilized eggs arrest at early stages, the capacity for full development must require additional egg factors. This is an important point, because developmental biology literature is replete with references to *egg maturation*, which is used to describe simply the capacity to resume meiosis. The meaning would be clearer if the term egg maturation were reserved for those eggs with full developmental potential.

Therefore, factors within the ovary must actively maintain the germinal vesicle stage of the egg to allow full maturation of its cytoplasm and accumulation of stockpiles necessary not only for early development, but also for healthy fetal development. The identity of all the factors is not known, but the intracellular messenger, **cyclic adenosine monophosphate (cAMP)** plays an important role. Maintaining high concentrations of cAMP during culture of germinal vesicle stage eggs blocks the breakdown of the germinal vesicle and the resumption of meiosis. A full understanding of the factors that direct the egg's unusually

cAMP: a common abbreviation for cyclic adenosine monophosphate, a molecule formed in response to hormonal stimulation of cells; it is classified as a second messenger because it travels through the cell and activates kinases.

long arrest in early M phase could lead to new approaches for deriving stem cells for therapy.

As described previously, germinal vesicle stage eggs can be activated to resume meiosis simply by removing them from the ovary. In nature, as discussed in Chapter 3, the LH surge leads to the resumption of meiosis while the egg is still in the ovary. Therefore, it is assumed that the LH surge overcomes the ovarian blockade to meiosis, perhaps by stimulating the production of progesterone. This is an ongoing area of research. The egg is programmed to locate the metaphase I chromosomes near the periphery, close to the plasma membrane of the egg. This leads to unequal division of egg cytoplasm when the polar body extrudes with half the chromosomes, but only a tiny fraction of the cytoplasm. The remaining cytoplasmic stores are thus available to the activated egg.

The completion of meiosis I and the extrusion of the first polar body leads to chromosome rearrangement and the second arrest point for the egg at metaphase II. As discussed in Chapter 3, cMos is the kinase whose activity is essential for maintaining arrest at metaphase II. Since arrest at metaphase II is unique to eggs, it is not surprising that cMos is specifically expressed in eggs and not in other cells in the ovary. Moreover, it follows that cMos is not expressed in cycling somatic cells outside the ovary since metaphase arrests are rarely observed. Thus, cMos is an example of proteins whose expression is peculiar to eggs; they are part of the group of genes expressed in eggs collectively referred to as **maternal messages**. A full understanding of maternal messages and the controls on their deployment could lead to valuable new approaches for the derivation of stem cells from activated eggs.

Eggs undergo activation in a surprising variety of ways, described in this chapter as spontaneous egg activation, artificial egg activation, activation by sperm, and egg activation following nuclear transplantation.

Spontaneous Egg Activation

Since cMos plays such an essential role in maintaining metaphase II arrest, it follows that defects in cMos expression or activity could lead to spontaneous egg activation. Evidence that spontaneous activation actually occurs in nature stems from the formation of **dermoid cysts**, a relatively common occurrence in women. Available evidence suggests that failure of the follicle to rupture and release the egg at metaphase II can result in egg activation within the follicle. This gives rise to the cluster of cells characteristic of dermoid cysts. Further development can lead to **teratomas**, benign growths that include collections of differentiated cells that exhibit some characteristics of adult tissues, such as muscle and bone. As described in Chapter 3, mice genetically engineered without the cMos gene developed ovarian teratomas at a high rate.

This type of egg activation obviously does not involve fertilization by sperm and is termed **parthenogenesis**, from the Greek words parthenos, meaning virgin, and genesis, meaning birth. This naturally occurring phenomenon emphasizes that sperm are not required for egg activation, an observation further supported by many studies of parthenogenic activation of eggs of most species. Eggs activated without sperm are termed **parthenotes**. Parthenogenic activation of human eggs may lead to the derivation of lines of human stem cells, as has recently been shown for primate eggs (Chapter 1). For this to occur, several modifications of normal egg processes are necessary, as will be discussed in the following section on artificial egg activation.

Parthenote: a cleaving egg that was activated without being fertilized by sperm.

Release of stored calcium ions in precise pulses, like a metronome pacing the music, plays a major role in egg activation. The calcium ions serve as second messengers within the egg cytoplasm and in some way bring about the deployment of other stored components. The first wave participates in the completion of meiosis II, and the waves that follow initiate cell cleavages (Figure 4.1). Although many of the details are still unknown, it is clear that the calcium pulses must occur in carefully timed intervals for a period of several hours to ensure sustained activation of the egg. It is as if each pulse is responsible for activation of components whose activities must follow previous events and carry out functions essential to subsequent events.

One early event is the degradation of cMos, which allows meiosis to resume. Although evidence is limited, as suggested previously, eggs deficient in cMos may be more readily activated and less dependent on the strict regimen of calcium peaks. Under the usual rules of meiosis, half the metaphase II chromosomes are extruded as a second polar body, and the other half are enclosed in a nuclear membrane within the egg (Figure 4.2). This creates a haploid nucleus within the egg.

Alternatively, if formation of the second polar body is prevented, then all the metaphase II chromosomes become enclosed within the egg's nucleus and it becomes a diploid cell. Interestingly, a landmark paper written by John Eppig in 1977 describing ovarian teratomas in mice and humans reported genetic evidence that teratoma formation did occur after extrusion of the first polar body and that the resulting tumors were diploid, suggesting either that extrusion of the second polar body was suppressed or that the haploid set of chromosomes underwent DNA replication before activation and cell division initiated, resulting in the cluster of

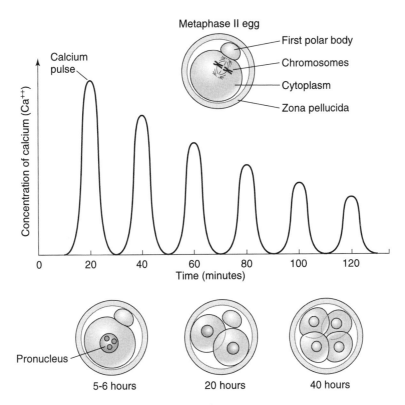

Figure 4.1   Release of calcium ions stimulates egg activation.

cells within the ovarian follicle. If the cells initiate regular cell cycles, then they may also respond to growth factors and differentiate accordingly, as described in Chapter 11. This could explain the formation of teratomas with several types of differentiated cells.

Artificial Egg Activation

One form of egg activation with the potential to give rise to two equal cells is to reprogram the germinal vesicle stage egg to bring about an equal cell division, yielding two "half-eggs" instead of one polar body and one egg (Figure 4.3). This experimental possibility has received little attention by developmental biologists, but could theoretically double the number of "eggs" available for stem cell production. Each half-egg would have a normal diploid number of chromosomes and theoretically could be activated to continue cell division parthenogenically. This possibility is an example of an area of research needed to improve the resources for stem cell derivation.

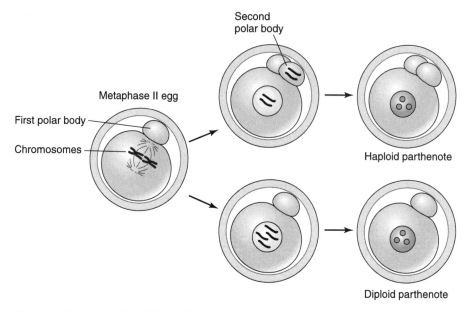

Figure 4.2 Haploid and diploid egg parthenotes.

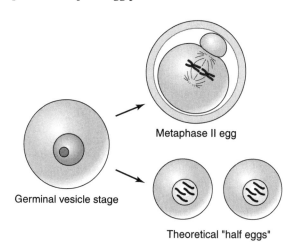

Figure 4.3 Theoretical splitting of the egg at meiosis I.

Development following parthenogenetic activation of eggs arrested at metaphase II, most frequently frog and mouse egg model systems, has been studied by numerous investigators for several decades. Egg activation without fertilization is regarded as a useful experimental tool with which to probe and understand those early developmental functions that are dependent on the molecular legacy of the egg. It also provides an opportunity to derive individuals that would be homozygous for maternal genetics, as will be discussed in Chapter 5.

Several types of chemicals can bring about activation of metaphase II eggs in culture, including molecules as simple as ethanol. Egg activation and one initial cleavage happens more readily in "aged" eggs than in freshly ovulated eggs, suggesting that cMos activity decays over time and cannot be renewed indefinitely. Aged eggs do not successfully sustain cleavage, however, even when activated by fertilization with sperm, suggesting that other maternal stores of essential cell cycle components may also decay.

Chemical activation includes compounds that inhibit protein synthesis, block microtubule assembly, stimulate release of calcium, and interact with egg plasma membrane receptors. Each species responds uniquely to such stimuli, supporting the concept that, although mammalian eggs share many features in common, details of developmental activation are species specific. Thus, the precise activation regimen that will lead to sustained cleavage of human eggs is an active area of research.

If the egg can undergo activation, cleavage, and differentiation of different cell types, the possibility exists that, under the right circumstances, offspring could form. There is no evidence for this in humans or other mammals, but this is a common occurrence among some insects, such as aphids. It is not known at this time whether mammalian eggs can be reprogrammed for full development of an offspring, or whether the requirement for a placenta limits developmental potential. It is clear that some genes are expressed differently if they are inherited from the father rather than from the mother. This suggests that the environment of the ovary and the testis influence specific genes differently, a process termed **imprinting**. Several characteristics of zygotes that relate to this question will be described in Chapter 5.

Imprinting: a modification to a gene that occurs in the ovary or the testis and influences what cells can express the gene without altering the base sequence of the gene.

Egg Activation by Sperm

Precisely how sperm contact activates the egg is an ongoing area of research. Animal models must be used principally in the United States because the U.S. Congress has decreed that, although research on fertilized human eggs "is meritorious and should be done for the good of society and humanity, the research will not be funded by taxpayer dollars." Therefore, the limited research that is done on fertilized human eggs in this country is done with private funds and without the benefit of organized oversight. The first reported attempts to fertilize human eggs in laboratory culture took place in Brookline, Massachusetts in the late 1930s and early 1940s (see sidebar).

Sperm entry into the egg leads to a cascade of events. Calcium spikes begin at regular intervals, thought to be mediated by factors brought into the egg by the

ACTIVATION OF HUMAN EGGS

In Vitro Fertilization and Cleavage
of Human Ovarian Eggs

Research that began in the late 1930s was partially reported in *Science* (1944;100: 105–107). The goal was to observe, in order to better understand fertilization and cleavage of human eggs. At the time the work was performed, fertilization of eggs by sperm, the fundamentally most important process for all species who sexually reproduce, had been observed in only a very few species. Dr. John Rock was an obstetrician/gynecologist at Harvard Medical School. His duties included caring for women at the Fertility Clinic Laboratory, Free Hospital for Women, in nearby Brookline. With the aid of his laboratory assistant, Ruth Menkin, they had recovered approximately 800 eggs from ovarian follicles of ovaries removed at hysterectomy. Of those, 138 were observed after exposure to sperm, and the 1944 report described the eggs from three women who underwent cleavage.

Mrs. DD: One cumulus-enclosed egg, recovered from a 2.3-cm follicle, was cultured in Mrs. DD's serum for 27 hours, then exposed to washed sperm for 1 hour in Locke's solution, during which time it was continuously viewed. "The spermatozoa showed great activity throughout the period of observation; they were clearly seen to travel through the interstices of the loose cellular formation surrounding the egg, and many were noted in active motion just outside the ovular boundary. At the end of one hour, the ovum was transferred to fresh serum from a post-menopausal patient." Forty hours later, it had cleaved to two blastomeres, each measuring 86 microns in diameter.

Mrs. RP: Thirteen eggs were recovered from developing follicles, cultured for 22 hours as before, exposed to sperm for 2 hours, transferred to post-menopausal serum, and observed 45 hours later. One (100 × 113 microns) had cleaved to two cells and had numerous sperm heads attached to the zona pellucida. The blastomeres measured 88 × 58 and 58 × 105 microns, respectively.

Mrs. JD: Four eggs were recovered from ovaries removed to seal the fimbriae because of "tuberculous endometritis." The eggs were cultured and observed as described previously. Two cleaved, one to three blastomeres. The egg measured 103 × 127 microns, the largest blastomered 97 × 73 microns and the two smaller ones were 62 × 62 and 50 × 63 microns, respectively.

These results, with respect to timing of cleavage, were in general agreement with the observations made on cleaving monkey eggs in 1933 by Lewis and Hartman (Carnegie Institute of Washington, Publication 443, contributions to *Embryology* 1933;24: 187). In all, four of the 18 eggs recovered from the ovaries of the three women exhibited one or two cleavages. Did Menkin and Rock observe the first fertilized human eggs or the first human parthenotes?

sperm itself. Such sperm factors have only recently been described and comprise an active area of research. The most striking distinction between egg activation following sperm penetration and artificial egg activation is the nature of the calcium spikes. Sperm penetration produces well-defined calcium spikes at precise intervals, sustained for the exact period of time necessary to support early embryonic events.

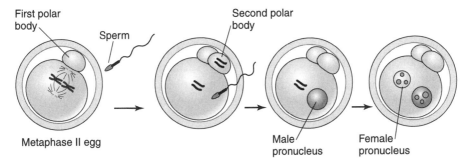

First polar body

Sperm

Second polar body

Metaphase II egg

Male pronucleus

Female pronucleus

Figure 4.4 Sperm penetration and formation of pronuclei.

Meiosis resumes at sperm entry and M phase is completed with the extrusion of the second polar body. Simultaneously, the sperm head is rapidly remodeled into a much larger, round nucleus by components in the egg cytoplasm (Figure 4.4). Sperm DNA is packaged tightly into the sperm head nucleus with unusual proteins, spermidine and protamine, which must be removed and replaced with standard somatic cell chromosome proteins, the histones. The remodeled sperm chromosomes are enclosed within a nuclear membrane and DNA replication begins almost immediately. Sperm do not contain the enzymes for DNA replication, so the DNA replication complex must be garnered from the egg cytoplasm. The sperm head remodeling process takes approximately 6 to 8 hours and is analogous to G1 in a somatic cell cycle (see Figure 2.8). While sperm head remodeling is occurring, the remaining egg chromosomes are enclosed within a second pronucleus, and replication of egg chromosomes also begins (Figure 4.4).

Pronucleus: the large nucleus with a few large, prominent nucleoli formed around the chromosomes of an activated egg and the sperm head, if fertilization occurred.

Because the sperm and egg chromosomes are individually haploid, the newly formed sperm nuclei are termed **pronuclei**. With a diameter of approximately 40 microns, the pronuclei are also extraordinarily large, leading to more open scaffolding of gene sequences than is present in somatic cells. The pronuclei are characterized by a few large nucleoli, and thus resemble chocolate chip cookies (Figure 4.4). The two pronuclei are unique to the zygotic stage, as discussed in Chapter 5.

Egg Activation Following Nuclear Transplantation

Given the ability of an activated egg to remodel sperm heads, it seemed likely that it could remodel somatic cell nuclei. Early interest in this process (described in more detail in Chapters 7 and 10) was for the purpose of testing the developmental potential of adult nuclei from frog somatic cells. After all, if all cells contain all genes, then it seemed theoretically possible that any cell in the body could direct the development of a new individual. Nearly 50 years ago, two groups of scientists reported the experimental results of nuclear transplantation into activated frog eggs. Nuclei were transferred into the egg cytoplasm by a variety of methods (Chapter 7). They discovered significant variations in the potential of nuclei to become remodeled by the activated egg. The more advanced the developmental stage of the transplanted nucleus, the less successful it was at supporting the development of a new embryo. Nonetheless, these experiments proved the principle that cells differentiated to become constituents of specific tissues and organs have retained all their genetic information.

Molecular Biology of Egg Activation

Meiosis. As discussed previously, one approach to doubling the number of eggs potentially available for stem cell derivation is to manipulate the cleavage at meiosis I into two equal cells, rather than the unequal division into a polar body and an egg arrested at metaphase II. Although this area has received little attention by developmental biologists, work to date indicates that an actin-rich region in the egg plasma membrane attracts and holds the first metaphase plate, and that if the actin filaments are disrupted or dispersed, the metaphase plate migrates back to the center of the egg. Therefore, it seems theoretically possible to engineer equal cleavage divisions of the germinal vesicle stage egg, giving rise to two half-eggs for progression to metaphase II (see Figure 4.3). There would likely be no reproductive advantage to the creation of half-eggs because of the limited developmental potential of the reduced cytoplasm, but there could be significant advantage to the creation of stem cells. The half-eggs could be activated either by parthenogenesis or following nuclear transplantation. This requires suppression of the pathways involved in maintaining metaphase arrest.

Zygote formation. Stimulating metaphase II-arrested eggs to resume development is first dependent on decreasing MPF activity, presumably by eliminating the actions of cMos. Since inhibitors of protein synthesis, such as cycloheximide, bring about activation, it is presumed that cMos is continuously translated during metaphase II arrest. Therefore, it is postulated that the first wave of calcium spikes brings about the degradation of cMos and the dephosphorylation of MAP kinase, leading to the resumption of meiosis (Figure 4.4).

Inositol 1,4,5-triphosphate (IP3) is an important cell-signaling molecule, termed a **second messenger** in response to a variety of stimuli. It provides a source of

Second messenger: a molecule created in response to a specific stimulus, such as hormone binding to a cell membrane receptor, which in turn brings about a specific cellular response such as activation of a kinase or a phosphatase.

phosphate groups for a number of kinases, and is formed in response to calcium ion oscillations brought about by a variety of stimuli. Some of its actions are mediated by binding to a receptor, termed IP3-receptor (IP3R). Molecules that bind to receptors are termed **ligands**. The IP3-IP3R ligand-receptor system appears to play a role in the resumption of meiosis and the successful suppression of MPF activity to allow subsequent cleavage divisions to initiate and continue.

The observation that elevations in cAMP concentration maintain arrest of the germinal vesicle suggests a role for protein kinase A, a principal target for cAMP. Protein kinase A phosphorylates a number of regulatory proteins, including some cell-cycle proteins (see Chapter 2).

Like "love's first kiss," sperm entry into the egg releases a protein that plays a key role in egg activation by mobilizing calcium. Sperm factor was originally described in 1996 and identified as a pentose phosphatase, but this is now known not to be its identity as a result of the work of several other laboratories. Identification and purification of this protein will pave the way to beginning to understand how the multiple calcium fluxes within the egg mobilize and deploy its stores of cell cycle components.

Additional Readings

Ducibella, T., Huneau, D., Angelichio, E., Xu, Z., Schultz, R. M., Kopf, G. S., Fissore, R., Madoux, S., and Ozil, J. P. (2002). Egg-to-embryo transition is driven by differential responses to Ca(2+) oscillation number. *Devel Biol* 250: 280–291.

Rosenblum, I. Y., and Heyner, S. (1989). *Growth Factors in Mammalian Development.* Boca Raton, FL: CRC Press.

Van Blerkom, J. (1994). *The Biological Basis of Early Reproductive Failure in the Human: Applications to Medically Assisted Conception.* Oxford: Oxford University Press.

Wu, H., He, C. L., and Fissore, R. A. (1997). Injection of a porcine sperm factor induce activation in mouse eggs. *Mol Reprod Devel* 46: 176–189.

CHAPTER 5

The Zygote

Make everything as simple as possible, but not simpler.

Albert Einstein

The Greek *zygous* means yoked and the French *zygous* means joined, terms that bring to mind teams of horses or oxen yoked together to share the workload. For centuries, the word has been applied to the most unique and unusual developmental stage of all individuals, the **zygote**. In the zygote, for the first time, genetic information from each parent is enclosed within separate, distinct, abnormally large nuclei, pronuclei, which form after fertilization. Toward the end of the zygote stage, the pronuclei come to sit adjacent to each other within the egg cytoplasm and together resemble the wooden yoke of the working team (Figure 5.1). This chapter reviews the zygote, including the requirements for both the maternal and paternal genes contained within the pronuclei to give rise to a new individual. In this way, the pronuclei are truly the working team the term suggests.

The Pronuclear Stage Egg

The human egg, like the eggs of most mammals, completes meiosis when metaphase II arrest is overcome by sperm entry. The metaphase II chromosomes resume meiosis almost immediately when the sperm enters, or the egg is otherwise activated, and the second polar body is extruded. Thus, human eggs do not naturally exist in a haploid state, as do sperm, and fertilization actually overlaps meiosis. Unlike the first polar body, the second polar body is haploid. The fate of the polar bodies is not known with certainty. The first polar body sometimes divides, sometimes disintegrates.

Figure 5.1 Human zygote.

There is no evidence that the second polar body divides; it is thought to disintegrate. There is renewed interest in polar bodies and their fate because of the possibility that analyzing their genetic composition may provide valuable clues about the genetics of the egg. Some couples undergo IVF so their embryos may be analyzed for specific, life-threatening disorders, such as Tay Sachs disease, known to be present in their genetic heritage. To avoid fertilizing genetically affected eggs, it is theoretically possible to analyze the polar bodies, and in this way surmise the genetics of the egg.

The process of sperm head remodeling begins immediately on sperm entry and lasts approximately 6 to 8 hours. While sperm head remodeling begins, the remaining egg chromosomes from meiosis II decondense into chromatin and become enclosed within a second pronucleus. As discussed in Chapter 4, this period is analogous to G1 in a somatic cell cycle (see Figure 2.8), except that the cellular machinery needed for DNA replication preexists in the egg cytoplasm.

The pronuclei have a diameter of approximately 40 microns, and each contains the haploid complement of sperm and egg genes, respectively. The reason for the extraordinarily large size of the pronucleus is not clear, but the circumstance leads to the possibility of more open scaffolding of chromosomes than is present in somatic cells. Such open scaffolding may ease constraints against gene expression, or DNA replication, as discussed at the end of this chapter. The pronuclei are characterized by a few large nucleoli, the organizing regions for ribosomal RNA, and resemble chocolate chip cookies (Figure 5.1). Pronuclei appear in human eggs approximately 6 to 8 hours after exposure to sperm and persist for approximately 10 to 12 hours, during which time DNA synthesis occurs within each separate pronucleus. Interestingly, several studies in the mouse have shown that DNA replication in the male pronucleus actually begins before the initiation of DNA replication in the female pronucleus. This suggests that sperm head remodeling may involve components of the DNA replication complex. This possibility may be important to successful nuclear remodeling, as discussed in Chapter 7.

Although not studied directly in human eggs, metaphase II mouse, rabbit, and goat eggs contain DNA polymerase activity at much higher levels than single somatic cells in S phase. This is reminiscent of frog eggs, which contain sufficient DNA polymerase activity to replicate all frog DNA within minutes instead of hours. But S phase in mouse and rabbit zygotes does not appear to be shortened by the increased amount of DNA polymerase activity. It requires the standard 6 to 8 hours observed for most mammalian cells. Therefore, since the increased amount of DNA polymerase activity does not apparently lead to shorter time for DNA replication, it may simply reflect sufficient stores to support more than one replication cycle without new enzyme synthesis.

The length of time it takes for DNA to replicate in fertilized human eggs is not known with precision, but available evidence suggests that it is comparable to the 6- to 8-hour DNA synthesis time it takes for human somatic cells. It is assumed, analogous to the mouse, rabbit, and goat, that unfertilized human eggs also contain extraordinary levels of DNA polymerase activity, as will be discussed in more detail in Chapter 6.

As described in Chapter 2, a fundamental feature of S phase is checkpoints to detect errors in DNA replication. The checkpoints allow the errors to be repaired before the cell cycle proceeds to cell division. Whether or not such a checkpoint exists in human zygotes is not known. The existence of such a checkpoint in somatic cells has been revealed by a variety of experiments, including purposefully damaging DNA during S phase (for example, by exposure to X-rays, or chemicals that damage DNA). Healthy cells will postpone the transition from S phase to G2 until the DNA damage is repaired, then proceed. Some studies also reveal increases in repair enzymes in response to DNA damage.

But zygotes rarely arrest in S phase, suggesting either that DNA repair enzymes are abundant and can repair damage quickly, or that there is no S phase checkpoint for DNA damage in zygotes. Since no single process is more fundamentally important to the development of offspring than faithful DNA replication in the zygote, such a checkpoint seems critically important. Nonrepaired errors in a DNA sequence would lead to damaged chromosomes in every embryonic cell. Clearly, defects in DNA replication and repair in zygotes could account for many of the failed embryos and is an area in urgent need of research.

Near the end of S phase, the pronuclei migrate to the center of the egg. The G2 phase is short, and M phase begins nearly concomitantly with the touching of the pronuclear membranes (Figure 5.2). Metaphase and anaphase within human eggs is less well studied than in animal species because of the U.S. Congressional moratorium against funding the research with taxpayer dollars. However, available evidence

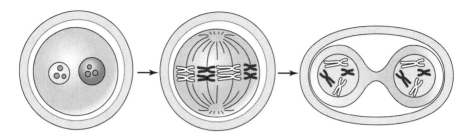

Figure 5.2 The first cleavage.

suggests that chromosome line-up and equal division into two daughter cells occurs in a fashion analogous to other cells. Cleavage proceeds immediately, delivering one chromatid from each parental chromosome into each daughter two cell. Since DNA replication is semiconservative, as discussed in Chapter 2, one DNA strand of each chromosome is the original strand from the sperm or egg, and the other strand is the one newly replicated in the zygote. The two-cell stage thus establishes for the first time a new genetic makeup within each nucleus.

In the mouse egg, a bridge between the two new daughter cells persists for a few hours after the cleavage furrow forms, allowing cytoplasmic contents to be shared between the two cells for a few hours. Unlike somatic cells, the first egg cleavage is not dependent on expression of genes in either the male or female pronucleus, although that is not known with certainty for human eggs. Nor is it known if the cell cycle machinery recruited for the first S, G2, and M phases is degraded after the first cleavage, as in somatic cells. What is known is that the G1 phase of the two-cell egg is short, suggesting that the egg's DNA replication machinery persists into at least the second S phase, as discussed in Chapter 6. This is consistent with direct measurements of DNA polymerase activity in mouse and rabbit blastomeres, which revealed relatively constant levels of enzyme activity during early cleavages.

Manipulation of the Zygote

The clearly visible pronuclei provide the opportunity for several types of manipulations of the zygote to explore fundamentally important questions about developmental potential. Probably the first zygotic manipulations were performed by Hans Spemann in the early 1900s. As will be described in more detail in Chapter 7, Spemann was able to manipulate the zygotic nuclei of a fertilized newt egg into one half of an egg that developed normally. The other half of the egg failed to undergo any development, demonstrating that the information guiding embryonic development resided in the nuclei.

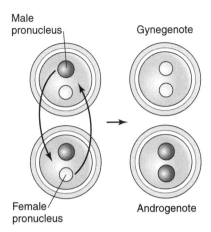

Figure 5.3 Exchange of pronuclei between zygotes.

Other types of manipulations of mouse zygotes revealed the importance to embryonic development of the presence of both the paternal and maternal genes. The pronucleus developed from the sperm head is larger and forms earlier than the female pronucleus. Several teams of investigators have removed one pronucleus from a zygote and replaced it with a pronucleus from another zygote. In this way, it is possible to reconstruct zygotes with two female pronuclei and/or two male pronuclei (Figure 5.3). Zygotes with two female pronuclei are termed **gynegenotes** and those with two male pronuclei are termed **androgenotes**. Such reconstructed mouse eggs have been used in several types of experiments designed to understand the respective contribution of the maternal and paternal genes to the development of offspring.

Gynegenote: a cleaving egg with two female pronuclei at the zygote stage.

Both gynegenotes and androgenotes undergo development to blastocysts with nearly the same frequency as bisexual zygotes. The resulting blastocysts also implant in the lining of the uterus (discussed in Chapter 8) and undergo early organ development at nearly normal rates, but neither type of unisexual embryo gives rise to live offspring. Gynegenotes exhibit relatively normal development of the embryo proper, but retarded development of extraembryonic membranes, including the placenta. Androgenotes exhibit retarded development of the embryo, but relatively normal development of extraembryonic membranes and placenta. Thus, unlike the oxen, the work of developing an offspring requires not just a cooperative pair, but one male and one female member to the team.

There are two possibilities for androgenotes because the zygote can contain either an X or a Y chromosome from the father. If it contains a Y chromosome, the offspring would be male; an X chromosome leads to a female. This suggests that differences in developmental potential could be influenced by the chromosome makeup of the androgenote, that is, XX, XY, or YY. In fact, the chromosome composition of the androgenotes had little influence on embryonic development; what was clearly important was the absence of the female pronucleus. This was a puzzling observation since the chromosome composition of a heterosexual zygote could also be either XX or XY, with the X chromosome always being derived from the mother, and the other X or Y chromosome being derived from the father. A homosexual gynegenote would always be XX. The difference in developmental potential of the zygotes thus appeared to result from the origin of the X chromosome: ovary or testis.

Several types of studies have since confirmed that the expression of a gene is not only determined by its position on the chromosome, but may also be influenced by whether it passed through an ovary or a testis. This type of genetic influence is termed imprinting. Although all the mechanisms of genetic imprinting are not known with certainty, one possibility is the addition of a methyl (–CH3) group to one of the DNA molecules (cytosine, described in Chapter 7). This process is termed **DNA methylation** and does not alter the fundamental sequence of the gene, but appears to influence whether or not the gene is expressed, and under what circumstances. Such modification of a gene is termed **epigenetic** to indicate that the gene sequence itself is not altered. Although there are some exceptions, as a rule regions of chromosomes whose genes are methylated on many cytosine residues (**hypermethylated**) are not expressed. Removing the methyl groups may allow the gene to become expressed. Several fundamentally important imprinted genes have been identified in the mouse.

> **Epigenetic:** a modification to a gene that does not involve changes in the sequence of the bases.

Other types of experiments have attempted to understand communication between the pronuclei and the cytoplasm. A particularly interesting experimental system is a mouse strain, DDK, that exhibits mating specificity. Females mated to mouse strains other than DDK males exhibit very low pregnancy rates, although their eggs undergo fertilization and one or two cleavage divisions. Other female strains mated to DDK males exhibit normal fertility. Series of reciprocal pronuclei crosses between strains has revealed there is a specific interaction between DDK male pronuclei and DDK egg cytoplasm that is essential for fetal development. Understanding the nature of this interaction may provide important clues about the continuing role of egg cytoplasmic stores in fetal development.

In summary, the zygote is the result of the remarkable ability of the egg to reorganize chromosomes into pronuclei. It seems likely that factors that can remodel both sperm heads and metaphase II chromosomes also can be used to remodel the nuclei from somatic cells, as will be discussed in Chapters 7 and 10. With respect to reproduction, parental DNAs are maintained as separate sets of chromosomes throughout the duration of the zygotic cell cycle, until the first metaphase occurs. It is the two-cell stage, not the zygote, that co-mingles parental chromosomes for the first time to define a new individual. Thus, "life begins at the two-cell stage" is more accurate than "life begins at fertilization." However, the paternal genes may not be expressed, and thus play no role in development at the two-cell stage. A better definition of "new life" may be when the paternal genes play a role in guiding development. This is not known with certainty for most species. The best available evidence suggests the paternal genes are expressed for the first time at the four- to eight-cell stage of human development, as will be discussed in Chapter 6.

Parthenotes

As discussed in Chapters 1 and 4, parthenogenetic activation of human eggs could be of particular value in the derivation of human embryonic stem cells. Two types of parthenotes have been studied, principally from mouse and rabbit eggs. One type is diploid and the other haploid (see Figure 4.1). Egg activation is brought about by a variety of external stimuli, including electrical impulses and chemicals termed **ionophores**, that stimulate release of calcium stores.

Haploid Parthenotes

A haploid parthenote results from egg activation with extrusion of the second polar body. Strictly speaking, this type of parthenote would be a haploid gynegenote. Studies of haploid mouse parthenotes revealed that early cleavage divisions can occur with a haploid number of chromosomes, indicating that the zygotic cell cycle does not depend on the full diploid complement of chromosomes. As discussed previously, this may be because checkpoints for DNA replication are not as rigorous in the zygote as they are in somatic cells. This circumstance may be necessary to allow entry into G2 and M phase following replication of the haploid number of chromosomes in each pronucleus.

Importantly, stem cell lines have been derived only from haploid parthenotes that have become diploid during subsequent cell divisions. For this to occur, a cycle of DNA replication with no cell division had to take place, presumably within the first few cleavages. Thus, the haploid nature of the genome may be sensed in subsequent cell cycles. If diploid parthenote stem cell lines could be derived from activated human eggs, it could provide a source of well-defined stem cells of a single set of genetics without the social stigma and more complex genetics associated with stem cells derived from human embryos.

TRANSGENIC MICE

Introduction of Foreign Genes into Zygotes

The advent of remote-controlled devices attached to microscope stages that could manipulate tools within a range of a few microns paved the way for new types of experiments in several fields of biology. Originally developed to inject compounds that could not pass through cell membranes directly into the cells, the devices were quickly adapted to embryo manipulations by developmental biologists.

Two groups of scientists applied this technology to mouse zygotes in the 1980s with the goal of understanding controls on gene expression in mice. Holding the zygote by mild suction with a polished glass tube approximately 30 microns in diameter (the *holding pipette*), the pronucleus of the mouse zygote presents a clearly visible target for another glass tube, pointed, sharpened, and less than 10 microns in diameter (the *injection pipette*). This *micromanipulation* system provides the opportunity to introduce substances, including DNA sequences, directly into the pronuclei of zygotes. The obvious question was, what would happen? Did the mouse embryos have the same "foreign DNA" sensing mechanisms as bacteria? If so, the injected sequences would be degraded. Amazingly, not only were the sequences not instantly degraded, but in many cases, they persisted into the offspring. This acceptance of new DNA and plasticity of the mammalian genome was unexpected.

In 1981, Ralph Brinster, working with Richard Palmiter, microinjected a DNA construct containing the gene for herpes virus thymidine kinase into one pronucleus of mouse zygotes, transferred them back into oviducts of female mice for development,

and demonstrated that not only had the DNA construct been carried forward into the developing mouse, but that some tissues expressed the thymidine kinase enzyme activity. This team followed this experiment in 1982 with a dramatic demonstration of the effects of injecting the gene for growth hormone into a mouse zygote. Of the 21 offspring that survived the procedure, seven, approximately 33%, carried the injected "transgene." That percentage of "transgenic mice" produced from microinjection experiments has remained consistent in subsequent experiments in many other laboratories. Of the seven animals that were transgenic for human growth hormone, six grew dramatically larger than their littermates.

In 1982, John Gordon and Frank Ruddle had also developed the techniques for DNA microinjection into mouse pronuclei. These investigators were the first to demonstrate that some of the transgenic animals could transmit the new genes to progeny, in what appeared to be standard Mendelian genetics. Animals with the new genes incorporated into either their oogonia or spermatogonia were termed founders.

That new genes could be so easily inserted into the germ line not only has far-reaching consequences for our views of evolution and inheritance, but provides an unprecedented and powerful opportunity to study controls on gene expression, as well as the influence of specific genes on cell functions. Thousands of transgenic mice have been created in laboratories throughout the world in the 20 years since the technology became available.

In 1989, an Italian team of investigators, headed up by Corrado Spadafora, demonstrated another approach to creating transgenic mice. They reported that exposure of sperm to solutions of DNA constructs allowed the incorporation of the DNA into the sperm head, and then into the egg at fertilization. This report could not immediately be confirmed by other scientists, including Ralph Brinster and Davor Solter, and doubt was cast on the reliability of the information reported by Spadafora's research team. However, several other teams of scientists throughout the world have now reported sperm-mediated DNA transfer in several species, including fish, amphibia, birds, and a few mammals. Creating transgenic mice by this approach remains more difficult than with other species for unknown reasons and is an active area of research.

Given the potential exposure of sperm to foreign DNAs in transit to fertilization of an egg, it is alarming to consider the possibility that those foreign DNAs could be incorporated into the human genome. This could hold particular relevance to programs of assisted reproduction throughout the world in which fertilization occurs under a wide variety of laboratory conditions.

Another type of haploid parthenote has been studied to determine the role of egg chromosomes in remodeling the sperm head. Activated mouse eggs with the egg metaphase plate removed (termed enucleated) have been injected with sperm. Remodeling the sperm head into a male pronucleus proceeded at a normal rate in the enucleated mouse eggs, confirming the notion that egg cytoplasm contains stockpiles of cell cycle factors needed for nuclear remodeling and DNA replication. Strictly speaking, injecting a sperm into an enucleated egg is not fertilization, because the egg genetic information is gone. Presumably, stem cell lines would be difficult to derive from such haploid androgenotes, because sperm contain either X or Y chromosomes, but not both. Thus, individual sperm do not represent the genetic composition of the man producing them.

Diploid Parthenotes

A diploid gynegenote parthenote arises from artificial egg activation before the second polar body is extruded, thus maintaining all 46 chromosomes of the metaphase II egg (see Figure 4.1). Under the right conditions, diploid gynegenotes can give rise to stem cells, as recently reported for eggs from the monkey, *Macaca fascicularis*. Primate parthenotes undergo developmental arrest in utero, as described previously for mouse zygotes, and are therefore not really embryos. Embryonic stem cells derived from diploid activated human eggs may, therefore, also be free from the social stigma associated with deriving stem cells from human eggs fertilized with sperm.

Diploid gynegenotes develop more reliably than do haploid parthenotes. Suppression of the extrusion of the second polar body by the activated egg is usually accomplished by culturing the egg in a common inhibitor of microtubule formation, as described below. In theory, it may also be possible to create a diploid androgenote from one X sperm and one Y sperm. Since this type of parthenote presumably would also not be capable of development, analogous to the mouse androgenotes, such androgenote parthenotes might also be more socially acceptable because they are not truly embryos. Work in this area is limited.

Once an interesting experimental tool for developmental biologists, parthenotes may become an especially valuable tool for stem cell medicine. Given the many ways that eggs may be activated without the potential to give rise to offspring, it becomes clear that new terms are needed to describe the processes. One principal difference between activating eggs to derive embryonic stem cells and activating eggs to give rise to an embryo is the absence of the creation of new germ cells with new genetics. Thus, the eggs are being activated to give rise only to new somatic cells to treat the diseases outlined in Chapter 12. A new term, **ovasome**, has been proposed to replace the term **embryo** when the goal of activating the eggs (**ova**) is to create new somatic cells ("somes"). The process, whether by parthenogenesis or nuclear transplantation, to be described in Chapter 7, could be termed **ovasomagenesis**.

> **Ovasome:** a new term to describe egg activation for the purpose of creating stem cells rather than an embryo.

Molecular Biology of Zygotes

Once the pronuclei have formed, the molecular pathways that controlled the arrest of the egg either at the germinal vesicle stage or in metaphase II have been eliminated. In addition, the cytoplasmic factors capable of remodeling the sperm head and forming nuclei have also been eliminated. Several investigators have discovered that either sperm or somatic cell nuclei, injected into zygote-stage eggs, do not undergo the nuclear remodeling characteristic of eggs within 1 to 2 hours of activation. Thus, those components in the activated egg responsible for nuclear remodeling are degraded within a few hours. It will be important to nuclear transplant stem cell derivation to fully understand those egg factors responsible for nuclear remodeling, as discussed in Chapter 7.

The giant cell enters the G1 to S transition, with DNA synthesis initiated independently in both pronuclei. Although a few other cells (e.g., placenta trophoblast, discussed in Chapter 8) contain multiple nuclei, they are formed by nuclear division without cell division. The need to coordinate DNA replication in two nuclei simultaneously is peculiar to the zygote.

Since, in general, progression through the zygote stage is blocked in the presence of inhibitors of protein synthesis, but not inhibitors of messenger RNA synthesis, it is assumed that protein synthesis, perhaps the cyclins, are required, and that their synthesis is guided by messenger RNA stockpiles within the egg. There is some evidence that paternal genes may be transcribed during the zygote stage, but they do not appear to be essential for cleavage to the two-cell stage.

As discussed previously, the G2 phase in the zygote is short. Once the pronuclei meet in the center of the egg, the G2 to M transition commences. An area of fundamental importance to the first cleavage is organization of the **centrosome**, the cytoplasmic organizer that defines the cleavage poles of the egg. In the mouse, the egg contains two centrosomes, but in the human, it is the sperm that brings the centrosome into the egg. Thus, a challenge to nuclear transplantation studies in human eggs is how to bring about successful cell cleavage in the absence of sperm centrosomes. This is an area of active investigation.

A peculiar characteristic of zygotes, including mouse and human, is their dependence on exogenous pyruvate for cleavage. This fundamental observation was made by Roy Whitten in 1957 as he attempted to define laboratory culture conditions that would support early development of mouse and rabbit embryos. He had observed development to blastocysts in culture by two-cell mouse embryos, but zygotes were blocked before cleavage. The inclusion of pyruvate in the culture medium solved the problem. More recent work has shown that pyruvate is necessary to support protein synthesis in the zygote, supporting the concept that protein, but not messenger RNA, synthesis is necessary for progression through the first cell cycle. Many of the metabolic pathways of zygotes and blastomeres have not been described, but the requirement for pyruvate, and the lack of utilization of glucose for energy, may relate to the quiescent nature of oocyte mitochondria.

Additional Readings

Wassarman, P. M., ed. (1991). *Elements of Mammalian Fertilization*, Vols. I and II. Boca Raton, FL: CRC Press.

Monk, M., ed. (1987). *Mammalian Development: A Practical Approach*. Oxford: IRL Press.

CHAPTER 6

Blastomere Cleavage

We shall not cease from exploration and the end of all our exploring will be to arrive where we started and know the place for the first time.

T. S. Eliot

 OVERVIEW

Once activated, the egg initiates a series of unique cell cycles, each of which results in identical daughter cells with smaller volumes. The reduction in size is the reason these cell cycles are termed **cleavages** rather than cell divisions. Before programs of human in vitro fertilization became standard of care for infertile couples, the term embryo was not applied to early cleavage stages. These stages were referred to by developmental biologists as *cleaving eggs* (sidebar, Chapter 4). The term cleaving eggs was confusing to infertility patients, however, because it did not adequately describe whether or not the egg had been successfully fertilized before it started to cleave. Thus, the term embryo became commonly used to refer to cleaving eggs that had exhibited the zygote stage following exposure to sperm.

The use of the term embryo for these early egg cleavage stages has led to substantial confusion, however, about the developmental potential of the fertilized egg. Moreover, the term embryo suggests that the newly compiled complement of genes are controlling early cleavage events. This is not the case. Several lines of evidence make it clear that the egg is responsible for guiding early cleavages. For example, they take place with or without the process of fertilization, as described in Chapters 4 and 5. Exactly when the new complement of genes takes over control of each cell is not known with certainty. It appears most likely to be a somewhat gradual process that occurs over several cell cycles. For accuracy, the term embryo should only be applied to those developmental stages in which the newly compiled embryonic genes are controlling cell functions. Those individuals wishing to define precisely when a new life begins should perhaps reserve that definition for the "embryonic **genome** control" stage, which is not known with certainty for human eggs at this time. Additional research is needed.

> **Genome:** a general term for all the genetic information of an organism.

The First Cleavage

The first cleavage occurs approximately 22 and 26 hours after exposure of mouse and human eggs to sperm, respectively. Parthenogenic stimulation leads to approximately the same time course.

However activated, the cleaved human egg gives rise to two smaller cells approximately 80 **microns** in diameter, each comprising approximately 270,000 cubic microns. Careful measurements of mouse embryos have revealed relatively little change in mass with respect to overall weight, protein, lipid, and nucleic acid content throughout early cleavages. Assuming the same for human eggs, the contents of 800,000 cubic microns in the human egg (Chapter 3) are therefore divided between two cells with a combined volume less than that, resulting in an increase in concentration of cytoplasmic contents. In contrast, the surface area of the egg, approximately 42,000 square microns, is divided approximately evenly between the two blastomeres, each with a surface area of approximately 20,000 square microns. The net effect of the first cleavage is, therefore, an increase in the ratio of plasma membrane to cytoplasm (Table 6.1) whose components, including signal transduction pathways, have undergone concentration. Whether or not this increased ratio of cell surface to cytoplasm influences the blastomere's response to its environment through cell surface sensors is a matter of speculation.

As mentioned in Chapter 4, a unique physiologic requirement of the zygote to undergo the first cleavage is the presence of pyruvate instead of glucose for energy. This fact was discovered by Roy Whitten, a New Zealand scientist, while trying to understand why two-cell stage eggs recovered from mice for culture in the laboratory would develop for four more days, to the blastocyst stage, whereas, one-cell stage eggs frequently arrested in the same culture conditions. In an effort to measure culture medium components with as much precision as possible, Whitten turned to calcium pyruvate as a calcium salt, which did not spontaneously attract as much water during storage as did other calcium salts, such as calcium chloride. Amazingly, in this new medium, zygotes cleaved to two cells, and then on to four cells, thus establishing

| | | Volume | Surface Area | Ratio of |
| | Diameter | (cubic | (square | Surface |
Stage	(microns)	microns)	microns)	Area/Vol
Egg	115	800,000	42,000	0.05
Two-Cell	80	270,000	20,000	0.07
Four-Cell	60	112,000	11,000	0.1
Eight-Cell	42	39,000	5,500	0.14
Sixteen-Cell	30	14,000	2,750	0.2
Thirty-two-Cell	21	5,000	1,400	0.28

Table 6.1

Physical Dimensions of Early Cleavage Stages

that, unlike most somatic cells, pyruvate is essential for the first cleavage division. In most cells, glucose is metabolized to pyruvate in the cell cytoplasm, and the resulting pyruvate is used to generate energy (Figure 6.1). The zygote and early cleavage blastomeres may, therefore, not contain the enzyme pathways for the initial steps of glucose metabolism. Since one of the first steps in glucose metabolism requires more energy than it creates, perhaps there is a developmental advantage to not using precious energy stores in the egg for glucose metabolism. These considerations are fundamentally important to the design of laboratory conditions to support egg activation and early cleavages, as is discussed later in this chapter.

The Second Cleavage

The two-cell stage (hereafter referred to as Two-Cell) marks the first co-mingling of maternal and paternal genes. In theory, the chromosome separation at the first cleavage event resulted in one maternal and one paternal member of each chromosome pair being allocated to each blastomere. As discussed previously, given the semiconservative replication of DNA, each new chromosome created in the zygote has one germ-cell strand of DNA and one zygote strand of DNA (see Figure 2.6).

The G1 phase of the Two-Cell blastomere is short for both mice and human eggs, and S phase begins within a couple of hours of cell cleavage. Interestingly, but for reasons that are not clear, the mouse Two-Cell maintains a cytoplasmic bridge between the two blastomeres until shortly after S phase begins. Information is lacking about whether or not this situation is true for human Two-Cells. The short G1 phase provides support for the concept that precursors for the DNA replication

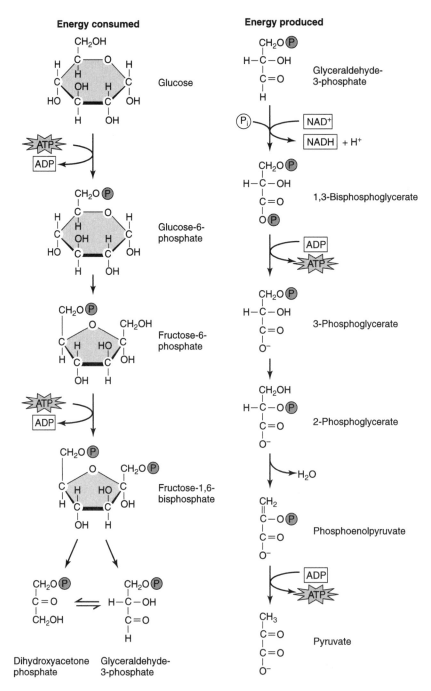

Figure 6.1 Glucose metabolism.

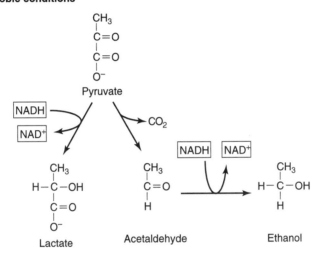

Figure 6.1 (*Continued*).

machinery preexist in the Two-Cell. The checkpoint for "start" may not exist, although cyclin synthesis and degradation seems to occur. S phase requires 6 to 8 hours for completion, in keeping with the S phase of somatic cells.

Half of the chromosomes resulting from S phase in the Two-Cell are comprised of double-stranded DNA in which both strands have been replicated after egg activation. The other half of the chromosomes contain one new DNA strand and one template DNA strand contributed by the egg and/or sperm. As discussed in Chapter 5, imprinted genes are thought to be controlled by methylation patterns, so called epigenetic regulation. Precisely how methylation patterns are reestablished in the newly synthesized DNA strand is not known with certainty, although there is evidence that the enzymes responsible for adding the methyl group, "DNA methylases," follow the pattern present on the opposite strand. Such epigenetic controls on gene expression are an active area of ongoing research.

The G2 phase of the mouse Two-Cell is unusually long, about 12 hours. The reasons for this are unclear, although many processes are known to occur during this phase, as discussed in the Molecular Biology section of this chapter. The long G2 of the mouse Two-Cell is also a period during which developmental arrest may occur, termed the *two-cell block*. A number of factors contribute to the two-cell block, including the genetics of the egg and the presence of glucose and some purines. A full understanding of this unusual cell cycle arrest will provide important insights into controls on blastomere cleavage.

Spontaneous separation of the two cells is one mechanism of twinning, including approximately one third of identical human twins. This observation speaks against the notion that "life begins at fertilization," because in the case of the identical twins resulting from separation of early blastomeres, their individual lives obviously do not begin until that separation occurs. This observation also emphasizes an important distinction between mammalian embryos and those of lower animals, namely, that each early blastomere maintains totipotency to develop into all tissues necessary for gestation and development into an offspring, including placenta and extra-embryonic membranes. This is in contrast to insects and amphibians whose eggs are polarized and where distinct regions of the cytoplasm are precommitted to becoming head, tail, back, and belly. In contrast, therefore, mammalian embryos must rely on other signaling mechanisms to bring about cell commitment events that will lead to tissue and fetal development. Although this is true, more recent evidence suggests that factors within the egg itself actually continue to influence fetal development. Unraveling these mysterious pathways is an urgently needed area of research, especially for human embryos.

Unlike the mouse, the G2 phase of the human Two-Cell is not unusually long, and cleavage to four cells occurs within approximately 12 to 18 hours (Figure 6.2). There is, however, a wide variation of cell cycle times for the early cleavages in the human, which is thought to relate to developmental potential. Many reports from infertility programs have revealed that pregnancy rates are substantially higher if the more rapidly cleaving embryos are selected for transfer into the uterus of the woman undergoing IVF. This has led to the notion that the developmental potential of more slowly cleaving embryos is reduced relative to the faster cleaving embryos, but many other variables could account for the variation in pregnancy rate, such as asynchrony of slower cleaving embryos with the uterus so that development is out of phase for the uterine lining (discussed in Chapter 8). Until testable parameters are established to reliably determine developmental potential, this will remain an area of speculation.

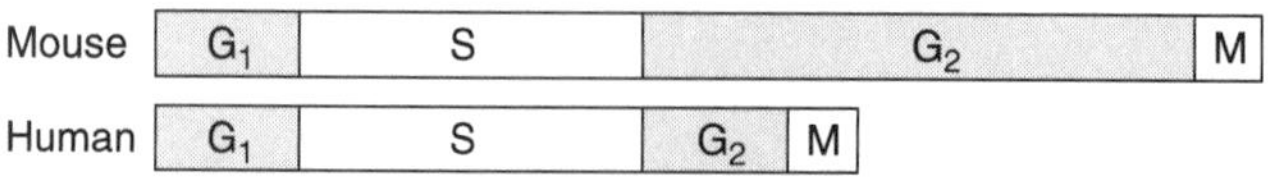

Mouse	G_1	S		G_2		M

Human	G_1	S	G_2	M

Figure 6.2 Cell cycle of mouse Two-Cell and human Two-Cell.

Cleavage of the Two-Cell to four cells occurs relatively synchronously, usually within a few minutes of each other for mouse embryos. Human Two-Cells are somewhat less synchronous, however, as evidenced by the relatively common appearance of Three-Cells. Each Four-Cell human blastomere is approximately 60 microns in diameter, with a volume of approximately 112,000 cubic microns, slightly less than half that of each Two-Cell blastomere and approximately 1/7 the volume of the egg (Table 6.1). The surface area of each human four-cell is approximately 11,000 square microns, on the order of half that of the Two-Cell. Thus, the second cleavage also results in an increased concentration of cytoplasmic components, as well as an increase in the ratio of cell surface to cytoplasm. These considerations suggest a conservation and reuse of plasma membrane with each cleavage division, along with increasing concentrations of cytoplasmic constituents. This strategy could serve to limit the need for synthesis of new components for the daughter blastomeres. Compacting the constituents in the cytoplasm could compensate for losses that may occur with each cleavage. For example, nucleosides and their phosphorylated forms essential for DNA synthesis have optimal concentrations for enzyme activity. Decreasing the volume of the cytoplasm could serve to maintain optimal concentrations of nucleosides in the face of stores depleted by each round of DNA replication.

Fragmented Cleavages

An unusual difference between mouse and human blastomeres is the incidence of **fragmentation** in human blastomeres. The term fragmentation is used to describe the appearance of small blebs of plasma membrane-enclosed cytoplasm that are associated with early cleavage stages. It is as if part of a blastomere forms its own cleavage plane independent of the chromosomal alignment (Figure 6.3). The appearance of such fragments is relatively common in cleaving human embryos, and relatively rare in cleaving mouse embryos.

The relative differences in fragmentation between the two species may be related to an interesting difference between the sperm of the two species. As described in Chapter 2, cell division is oriented according to the positioning of the centrosome, a cytoplasmic structure that contains the **centriole** that is the organizing center for the formation of the mitotic spindle. Interestingly, mouse eggs contain their own centrosome, but human eggs do not. Unlike mouse sperm, which lack a centrosome, human sperm contain two. It is speculated that abnormalities in sperm centrosomes could account for some human fertility problems and aberrant cleavages of fertilized eggs.

Recent time lapse videos of cleaving mouse eggs, however, suggest another mechanism may also be operational (Figure 6.4). A cleaving Two-Cell was observed to undergo fragmentation followed by reannealing of the fragments. This observation emphasizes the plasticity of the membrane of the cleaving blastomere. That the

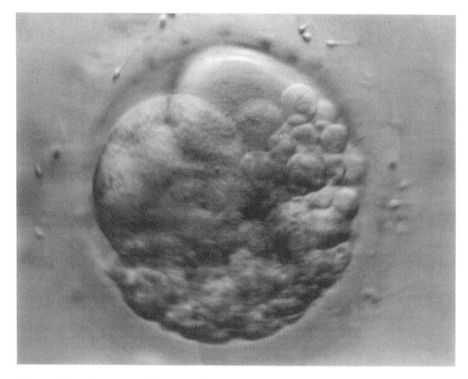

Figure 6.3 Fragmented human blastomeres.

human sperm centrosome is not essential for early cleavage events is, however, supported by several lines of evidence that parthenogenesis of human eggs does occur, both spontaneously in the ovary, as described in Chapter 3, and in laboratory conditions, as recently reported (Chapter 1). It will be important to determine if parthenogenically activated human eggs exhibit a greater incidence of fragmentation during cleavage than eggs fertilized by sperm.

The Third, Fourth, and Fifth Cleavages

Analogous to the Two-Cell mouse, it is the human Four-Cell that is prone to developmental arrest (Figure 6.5). There is no evidence that the G2 phase of the human Four-Cell is as unusually long as the mouse Two-Cell, but arrest at the four-cell stage is commonly reported by infertility clinics. This may be related to the timing of the onset of blastomere gene expression, as described in the Molecular Biology section of this chapter.

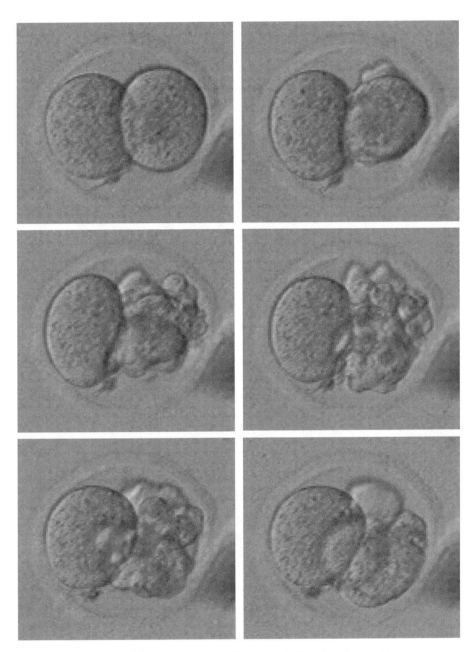

Figure 6.4 Phases of blastomere reannealing recorded by time-lapse videomicroscopy. The time between frames is approximately 7 minutes.

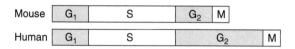

Figure 6.5 Cell cycle of mouse Four-Cell and human Four-Cell.

The blastomeres that do not undergo cell cycle arrest cleave again within 18 to 24 hours. The third cleavage heralds the onset of the **morula** stage of development and is not as synchronous as the second cleavage. Assuming that the paradigm that plasma membrane will be completely recycled persists with each cleavage, the eight-cell blastomere will have a surface area of approximately 5,500 square microns, a diameter of approximately 42 microns, and a volume of approximately 39,000 cubic microns (Table 6.1). The 32-cell blastomere is approximately the size of a human somatic cell (Table 6.1), and as such is nearly 1/200th the size of the egg. Assuming no loss of egg cytoplasmic organelles, they have become concentrated more than fivefold. It is at this stage that each cell must begin to grow in size before division takes place.

> **Morula:** a cleaving egg with more than four blastomeres that has not developed a blastocoel cavity; usually 8 to 32 cells.

It is interesting to speculate that the strategy behind the enormous size of a mammalian egg is to support approximately 16 replications of the chromosomes before each cell must assume responsibility for continuing cell division. Since each cell remains totipotent, this paradigm allows for the loss of some blastomeres, perhaps because of errors in DNA synthesis, without jeopardizing the formation of a complete individual. Several lines of evidence have shown that removal of a few blastomeres at the eight- to 16-cell stage (**embryo biopsy**, Figure 6.6) does not compromise the developmental potential of the embryo. It thus seems reasonable to view the egg as an enormous, self-contained DNA replication factory, with the corollary abilities to remodel the sperm head and muster cell cycle machinery.

The notion that the egg is designed to be self contained for several cycles of DNA replication and cell division suggests it does not need the exogenous growth signals associated with progression through G1 and into S phase, as described in Chapter 2. Moreover, the well-documented toxicity of serum to early stage embryos in culture, particularly the mouse, suggests that growth factors may, in fact, derail the well-orchestrated egg programs for progression through early blastomere stages. In this regard, it is important to note that anatomically, these stages of embryonic development occur in separate compartments of the female reproductive tract. All well-studied mammalian species, for example, mouse, rabbit, cow, goat, sheep, hamster, and human, have oviducts (also termed fallopian tubes), which link the uterus to the ovary. The transit time through the oviduct coincides approximately

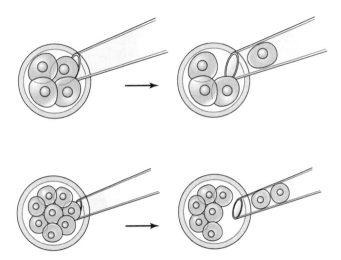

Figure 6.6   Biopsy of four-cell and eight-cell embryos.

with the stages of early cleavage divisions. Most embryos enter the uterus at the morula stage, as discussed further in Chapter 8. These characteristics hold important clues for the design of laboratory conditions to sustain these early cleavages, as discussed in the following section.

Embryo Development in Laboratory Conditions

The ability to grow cells in the laboratory (termed **in vitro**, meaning *in glass*) was recognized over a hundred years ago as essential to understanding the nature of cells and their metabolic requirements. Louis Pasteur proved that round cake-pan shaped dishes with loose fitting lids that overlapped the sides of the dishes could be used to culture bacteria in warm incubators without the bacteria escaping from one dish into another. Termed **Petri dishes** or **Petri plates**, they were modified to have optically perfect bottoms to allow the observation of cells placed into the dishes through "inverted" microscopes. Like bacteria, the cells also cannot escape from one dish to the other. The observation of living cells through the bottom of their culture dishes remains standard laboratory practice and has allowed the development of techniques for manipulating individual cells, as described in Chapter 7. The next problem was to understand what laboratory conditions would best support cell growth and division.

Maintaining a balance of acids to bases in blood is critical to sustaining life. This is because enzyme reactions require strict adherence to a narrow window of balanced

$$H_2O + Na^{++}HCO_3^{=} \rightleftarrows CO_2\uparrow + H_2O + NaOH$$

Figure 6.7 Bicarbonate buffering system.

acid/base conditions, measured as the concentration of hydrogen ions in blood. The measurement is performed with specially designed probes sensitive to hydrogen ions. The most commonly used term to describe the concentration of hydrogen ions is **pH**, a term related to the logarithm of the hydrogen ion concentration. Normal, physiologic pH values for blood are 7.2 to 7.4. To maintain this narrow range of pH, blood and other body fluids contain molecules that can acquire or release hydrogen ions as needed to maintain the balance. Such molecules are termed **buffers**, an important example of which is the **bicarbonate** buffer system. Carbon dioxide is a gas that is a little bit soluble in water. If added to water under pressure, once the pressure is released, the gas bubbles reform. This accounts for the bubbles in all *sodas*, a term derived from **sodium bicarbonate**. In solution, carbon dioxide can combine with water to form carbonic acid, which readily ionizes in the presence of sodium chloride (the most common salt in blood) to form sodium bicarbonate (Figure 6.7).

> **pH:** denotes the negative of the log of the molar concentration of [H$^+$] in a solution; pH 7 indicates [H$^+$] concentration equal to 10^{-7} molar.

Carbon dioxide is a byproduct of many normal physiological reactions, such as the metabolism of glucose (see Figure 6.1). The cells in the lung are designed to release the carbon dioxide into the air when we exhale. Exactly how much is released and how much is retained is determined by the pH in the blood at the moment of exhalation. The monitoring system is highly dynamic. A normal concentration of bicarbonate in the blood is 23 to 25 millimolar. In order to maintain a pH of 7.4 in blood, it is necessary to have a concentration of CO_2 equivalent to 6.5% of the air in the lungs. Since the concentration of CO_2 in the atmosphere is on the order of 0.2%, it is clear that inhaling does not provide much buffering capacity. How, then, can one maintain a normal, physiologic pH for cells cultured in the laboratory? Once the need was recognized, chambers were designed with carbon dioxide flowing into them in measured amounts mixed with air. Culture fluids (termed **culture media**) were formulated with 23 to 25 mM bicarbonate to mimic as closely as possible the physiologic conditions provided by the blood.

Blood also contains a well-defined mixture of salts and other ions. The salt concentration in blood is termed **normal saline**, and is responsible for maintaining the balance of water inside and outside cells. Cells placed in distilled water, for example, will swell until they burst because water will flow into the cell, attracted by the intracellular salt. It is not surprising, therefore, that the basis of all culture media is a salt solution very similar to that found in blood.

In addition, enormous efforts have been expended in defining and redefining the other ingredients in culture medium that are necessary to support many generations of identical cells. Starting with a single cell and allowing it to divide repeatedly is true "cloning" and can lead to the generation of millions of genetically identical daughter cells. This is clearly a more ideal experimental system than beginning with a heterogeneous population of cells from, for example, disaggregated tissues.

Initial studies to define the laboratory conditions that would support cell division revealed that a characteristic fundamental to cancerous tumors is that the cells grow readily in culture in a variety of conditions. More recent results have revealed that this is due to mutations in cell cycle checkpoint genes that allow cancer cells to escape normal cellular controls. For this reason, cancer cells in laboratory culture have escaped the routine cell cycle checkpoints for errors in DNA replication, as described for mutated p53 proteins in Chapter 2. Laboratory lines of cancer cells accumulate errors in DNA and even gain and lose chromosomes, a condition referred to as **heteroploidy**.

Heteroploidy: a non-characteristic number of chromosomes, or parts of chromosomes, in a cell.

Such cells are useful for some types of experiments, but not others. To study specific types of normal cells, scientists began to try to understand the laboratory culture conditions necessary to support cell division of single, normal cells. An important discovery in this regard was reported in 1948 by Dr. Katherine Sanford working in the laboratory of Dr. Wilton Earle at the National Cancer Institute (see sidebar). Dr. Sanford determined that single cells could continue dividing if they were cultured in medium "conditioned" by groups of cells, such as sections of whole tissue. In addition, by culturing single cells in small capillary tubes in a small amount of medium, she demonstrated that cells could condition their own medium if the volume were small enough. Not only did these experiments pave the way for expansion of normal cells important for many types of biomedical studies, but they suggested the existence of factors, produced by all cells, that stimulated cells to divide, so called **growth factors**. The need for such growth factors was tested by culturing normal cell lines on top of **feeder layers** of cells that were living but could not divide anymore because their DNA had been extensively damaged by X-ray treatment. Such conditioned medium was analyzed by numerous laboratories to better understand growth factors, of which many families have now been identified. As will be described in Chapter 9, human embryonic stem cells have been derived by culture on feeder layers with carefully defined growth factor supplements.

The search for the best culture medium for early cleaving embryos is an active area of research. Considering the autonomous characteristics of the blastomeres, culture needs for early stages may be relatively simple, designed to avoid premature stimulation of growth factor pathways and maintenance of the internal stores of precursors for protein and nucleic acid synthesis. Pyruvate and/or lactate are the most striking single requirements known for early cleavages.

DEVELOPMENT OF TISSUE CULTURE MEDIA

Developing the formulations of buffers, salts, vitamins, and other nutrients that supported cell division in mammalian cells was the work of many investigators. Early work used relatively high concentrations of serum (30% of the volume), usually recovered from fetal calves at slaughter houses because it was naturally sterile, and the remaining 70% was supplemented with a wide variety of additives based on the composition of various body fluids. The systems produced wide variations in results and yielded little information about fundamentally important requirements for dividing cells. Two milestone papers published in 1955 changed the field radically. One was published by Dr. Theodore Puck and Philip Marcus at the University of Colorado in Denver, entitled "A Rapid Method for Viable Cell Titration and Clone Production with HeLa Cells in Tissue Culture." The goal was to culture HeLa, the cervical carcinoma cell line used in many studies, in a completely chemically defined medium like bacteria. The other study was published by Dr. Harry Eagle at the National Institutes of Health in Bethesda, Maryland, entitled "Nutritional Needs of Mammalian Cells in Tissue Culture." Together, these works laid the basis for understanding the important roles of amino acids, vitamins, salts (including trace metals such as iron), glucose, and abundant small peptides such as

glutathione. Amino acids termed *essential* for survival were those the cells could not synthesize on their own; amino acids termed *nonessential* could also be synthesized by the cells. Those terms are still in use today as is the culture medium Basal Medium Eagle, BME.

Building on this basic medium formulation, numerous investigators have discovered specific requirements for specific cell types, thus yielding valuable information about the metabolic processes involved in various body tissues. Most hormones and growth factors have been discovered in this way.

Dr. Richard Ham joined the faculty at the University of Colorado to pursue his interest in culture requirements of cells growing individually instead of in large numbers. These studies have special relevance to cultured embryo cells, and it was Dr. Ham who formulated Ham's F-10 (he is now up to Ham's F-411) supplemented with pyruvate, which was necessary to promote cell division in Chinese hamster ovary cells, a cell line used in many experiments.

A few years later, Dr. Whitten would discover that pyruvate was essential for the first cleavage in the mouse zygote. Thus, it was not an accident that the first human IVF pregnancy, Louise Brown in England in 1980, resulted from fertilization of Mrs. Brown's egg in Ham's F-10.

Nonetheless, embryonic development in all available culture systems is not normal. This fact is well illustrated by mouse embryos. Because laboratory mice respond in a predictable fashion to exogenous administration of gonadotropins, development can be timed from the administration of the artificial LH surge (discussed in Chapter 3). Oocytes are released from ovarian follicles 12 hours after the injection of the hormone, and since the mice mate beforehand, the sperm are waiting in the oviduct.

Thus, in vivo development of mouse embryos is just as synchronous as development following fertilization in the laboratory. For this reason, the mouse represents a useful mammalian system for testing the safety of embryonic culture conditions.

A comparison of mouse embryos developing in culture with those developing in pregnant mice reveals a marked lengthening of the second and third cleavages. This leads to in vitro embryos with one half to one third the normal number of cells at the time the first cell differentiation occurs, as will be discussed in Chapter 8. Whether or not somatic cell nuclei transferred into activated eggs will require new formulations of culture medium for complete remodeling to a stem cell state awaits additional research.

Molecular Biology of Blastomere Cleavage

Sensitive techniques have detected mRNA synthesis in the G1 stage of the two-cell mouse embryo, but unlike somatic cells, the newly transcribed embryonic mRNA is not required for progression to S phase. In contrast, the mRNA synthesis that occurs during the long G2 phase of the Two-Cell is necessary for progression to four cells. The mRNA synthesis occurs even while oocyte mRNAs are being degraded. A peculiar feature of this cell cycle stage is profound degradation of residual oocyte mRNAs, as if in preparation for transition to embryonic genome control of cell functions.

Although it appears that the blastomeres do not require exogenous growth factors to proceed through the cell cycle, several lines of evidence suggest they may synthesize their own growth factors, a mechanism termed **autocrine** stimulation. One line of evidence is the need for the ras/raf pathway, a common signal transduction pathway that stimulates a critical cell cycle regulator, MAP kinase, as described in Chapter 2. This pathway is activated by a variety of growth factors acting on receptors at the plasma membrane of somatic cells. A blockade of the ras/raf pathway blocks two-cell mouse embryos at the two-cell stage. Either this pathway operates differently in blastomeres from all other cell types, or the blastomeres themselves secrete a growth factor that binds to its plasma membrane receptor and, in this way, activates the ras/raf pathway.

Autocrine: stimulation of a cell process by a signal (e.g., a hormone) produced by the cell itself.

In addition to the detection of cyclin and Cdc2, these observations indicate that the cell cycle controls on blastomere cleavage are similar to somatic cells, with the exception that preexisting stockpiles of maternal mRNAs and proteins are deployed rather than de novo synthesis of essential components. Studies of the DNA replication and repair machinery operational in blastomeres are limited. Faithful DNA replication is the single most important job of a blastomere. Understanding blastomere-specific mechanisms of DNA replication is urgently needed, both for the production of human embryonic stem cells, and for improving the safety and success of human IVF cycles.

Additional Readings

Katsuta, H. (1977). *Nutritional Requirements of Cultured Cells.* Baltimore: University Park Press.

Longo, F. J. (1987). *Fertilization.* London: Chapman and Hall.

Slack, J. M. W. (1983). *From Egg to Embryo: Determinative Events in Early Development.* London: Cambridge University Press.

CHAPTER 7

Early Nuclear Transfer Technology

What this will mostly be used for is to produce more health care products.

Ian Wilmut

OVERVIEW

Early reproductive biologists studied amphibian eggs since their large size allowed visualization with existing microscopes, termed dissecting microscopes, whose magnifying power was limited. The story is told (see sidebar) that it was Hans Spemann's baby's hair that was delicate enough to assist in the isolation of individual nuclei in early cleaving newt eggs. As a result of the manipulation of individual nuclei, Spemann reported in 1902 that embryonic salamander nuclei contained all the genetic information needed to guide the development of an entire salamander. This elegant observation contributed significantly to the concept that genetic information was encoded within DNA, the chemical structure of which would not be known for another 50 years.

Advances in microscopy, and the development of tools with finer control than a baby's hair, opened the way to ask questions about the genetic repertoire of more fully differentiated cells. A fundamentally important question was whether or not adult somatic cell nuclei had been genetically altered. Discovering the answer was important for many reasons. For example, if liver cells had lost genetic information, or duplicated existing genetic information to function as liver cells, then cell functions were probably not reversible (termed **dedifferentiation**), and the process of cell differentiation into specific tissues involved specific genetic alterations. Alternatively, if liver cell nuclei possessed all the information for a new organism, then the fact they were liver cells was not due to genetic alterations, but to epigenetic controls on gene expression. This information was fundamentally important to all of medicine, including developmental biology and cancer biology.

By the early 1940s, studies of frog embryo development had progressed to the discovery of laboratory methods for parthenogenic activation of frog eggs. This experimental system provided the opportunity to begin to explore the inherent developmental potential of nuclei from somatic cells.

Transfer of Nuclei into Frog Eggs

Two groups of investigators (see sidebar) studied nuclei from embryonic and adult frog cells throughout the 1950s. The results of their studies indicated that embryonic nuclei, but not nuclei from adult somatic cells, could direct development of new frogs. Following the results of those experiments, by the early 1960s, medical students and developmental biologists were being taught that embryonic nuclei retained all the information needed to direct the development of new frogs, but adult cell nuclei had been permanently altered and could no longer direct embryonic development.

The belief that the genetic information in adult somatic cells was permanently altered shaped scientific thinking for at least two decades. Similarities between the growth of embryonic tissues and the growth of cancerous tissues were obvious, but the concept that normal cells could dedifferentiate into cancerous cells was discounted because of the evidence that normal adult cell nuclei could not be reprogrammed to support embryonic development. Nonetheless, throughout this period, studies of DNA had revealed its basic structure, and numerous studies by many research teams of the DNA content of many types of cells revealed no gross differences between embryonic and adult cells with respect to chromosome number, DNA content, or DNA sequences. Other explanations for the inability of adult cell nuclei to support embryonic development were sought.

The frog egg experiments had demonstrated several important egg functions, including the fact that eggs did not need to be activated by sperm to remodel somatic cell nuclei. Remodeling chromatin is clearly what eggs are designed to do once they are activated, as described in Chapter 3. Moreover, nuclear remodeling and subsequent cleavages occur with or without the presence of the egg chromosomes.

DEVELOPMENTAL POTENTIAL OF TRANSPLANTED NUCLEI

Hans Spemann contracted tuberculosis in 1899, just after Mendel's 1866 paper on genetics had been "rediscovered." As a result of the discovery of Mendel's work, it had become broadly appreciated among biologists that genetic traits were inherited in a somewhat predictable mathematical fashion. During his recovery in a sanatorium, Spemann familiarized himself with the developmental biology theories being espoused by Weissman. At the time, it was understood that all individuals arose from a single fertilized egg that underwent multiple cleavages. Biologists were grappling with an explanation for cell differentiation—how did the profusion of different types of cells in each tissue happen? Weissman theorized that each time a cell divided, its chromosomes were split in half. Ultimately, the remaining chromosomes in a cell determined its function. Combining Mendel's and Weissman's theories meant believing that only sperm and eggs maintained the full complement of inherited traits. To Spemann's mind, this was a provocative possibility, but experimental data were lacking.

He lectured and studied at the University of Wurzburg while he worked toward his doctorate. He had chosen newts as his experimental system because the eggs were large and transparent, and the laboratory conditions for their development were known. His first experiment (using a loop of hair from his newborn son) was to separate the blastomeres of a salamander Two-Cell to determine if one developed into the head and the other into the tail, as would

be predicted by Weissman's theory. Instead, two complete salamanders developed. Ecstatic, he repeated the experiment by separating each blastomere of a Four-Cell. Four salamanders developed. Then eight salamanders developed from each blastomere of an Eight-Cell. However, after the eight-cell stage, development was erratic. He decided that at the 16-cell stage, the fates of each cell had been determined.

To test this theory, he coaxed both zygotic nuclei into one half of an egg and kept them there with another loop of his baby son's hair. After they had cleaved several times, he loosened the loop and manipulated one of the embryonic nuclei into the other side of the egg cytoplasm. This was the first nuclear transfer experiment. Amazingly, a second embryo formed. This proved that several nuclear divisions did not decrease the genetic information in the embryonic nucleus, and shot a big hole in Weismann's theory. Spemann's work was published in 1902.

Fifty years later, two groups of English scientists continued to explore the genetic repertoire of fully differentiated cells by transplanting nuclei into activated frog eggs. The first report was by King and Briggs in 1953. These investigators removed the genetic information from the eggs of the frog, *Rana pipiens*, and transferred in nuclei from cells of early Rana embryos and late Rana embryos. The early embryo nuclei directed the development to normal tadpoles in approximately one third of the nuclear transplants, but no tadpoles were obtained from transplanted nuclei of later embryos. A few years later, in 1958, Gurdon, Elsdale, and Fischberg performed similar experiments with another frog, *Xenopus laevis*. Their experiments differed in that they did not remove the genetic information from the eggs before transplanting nuclei from Xenopus embryos because they had a genetic marker to distinguish offspring resulting from the transplants. These investigators reported development to adults following nuclei transferred from developmental stages as late as a prehatching tadpole. Thus, activated frog eggs were shown to be able to remodel embryonic nuclei, but not adult cell nuclei, to give rise to new individuals.

These were the first "nuclear transfer clones," although neither group of scientists used the term "clone" to describe the tadpoles and frogs that developed.

Thus, the cellular components involved in nuclear remodeling and early embryonic development preexist in some form in the frog egg cytoplasm.

Given the wide variety of functions cells must perform, it seemed likely that some nuclei could be more easily reprogrammed than others. An important constraint on nuclear reprogramming by eggs is that it must take place simultaneously with egg activation to set up sustained cleavages. As described in Chapter 3, metaphase II-arrested eggs are in a carefully controlled cell cycle arrest. That arrest must be overcome to allow nuclear remodeling and cleavage. But the egg has a finite lifetime. Aged eggs are more readily activated, but fail to sustain as many cleavages as do "young" metaphase II eggs. Thus, the narrow time interval to accomplish nuclear remodeling and initiate successful cleavages may be insufficient for reprogramming some somatic cells. One possible solution to this problem was to allow the nucleus of a somatic cell to be remodeled in stages.

In 1970, Gurdon and Laskey tried this approach. They reported the development to adult frogs following a two-step transfer of nuclei originally taken from a monolayer of cultured epithelial frog cells (Figure 7.1). The epithelial cell nucleus was first

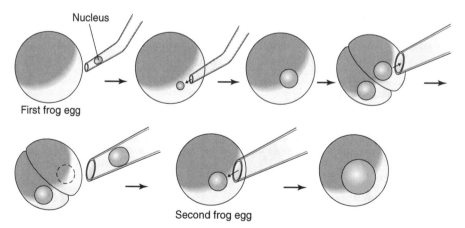

Nucleus

First frog egg

Second frog egg

Figure 7.1 Schematic of two-stage nuclear transfer experiments in frogs.

transferred into an activated egg, which then cleaved to an early embryo. A nucleus from the early embryo was then transferred into another activated egg from which the genetic material had been removed. A small percentage (less than 5%) developed through the tadpole stage to adult frogs. These results demonstrated that skin cell nuclei maintain the full genetic potential of an individual. The problem is in the nuclear remodeling.

Once Gurdon had established that frog eggs could remodel nuclei from adult frog cells, it was obvious to try to understand what would happen if nuclei from other species were transferred into frog eggs. The first cross-species human/frog nuclear transfers were reported in the late 1970s. The human nuclei were recovered from HeLa cells, a human cell line established in culture from a cervical carcinoma. HeLa cells have been used in many laboratories for a wide variety of experiments for several decades. They were an obvious choice for the experiments designed to see the effect of frog egg cytoplasm on gene expression by human cells.

Gene Expression: Transcription and Translation

The sensational human/frog experiments are actually built upon other work, a description of which will benefit from a brief review of the basic steps of gene expression. As discussed in Chapter 2, genes are unique and specific polymers of four molecules distinguished by side groups of adenosine (A), cytosine (C), guanosine (G), and thymidine (T). The arrangement of the molecules in the polymer determines the nature of the protein specified by the gene. Genes are joined end to end in the long strands of DNA that comprise the chromosomes. Genes are highly variable in length and are distinguished from each other by specific "start" and "stop" sequences (Figure 7.2).

As depicted in Figure 7.3, the expression of genes into new proteins requires a series of well-coordinated steps. First, the specific base sequence in one strand

Transcription: the process of synthesizing messenger RNA (mRNA) from the gene sequence as the template.

of the DNA that comprises the genes is **transcribed**, by a process termed **transcription**, into single-stranded RNA (ribonucleic acid) (Figure 7.4). Enzymes, termed RNA polymerases, actually synthesize the RNA to

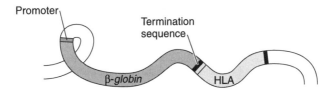

Figure 7.2   Schematic of genes arranged end-to-end on chromosomes.

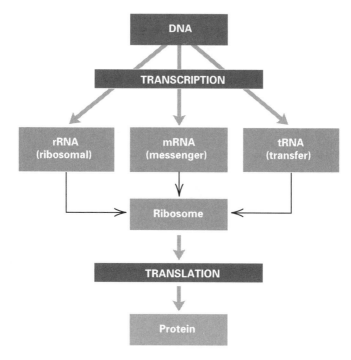

Figure 7.3 Flow scheme of transcription and translation.

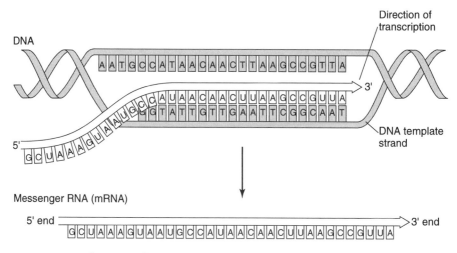

Figure 7.4 Schematic of transcription.

uniquely mimic the DNA sequences in the gene. The RNA strands contain molecules with three of the same four side groups (A, C, and G), and uridine (U) instead of thymidine. Three types of RNA are transcribed from individual genes: messenger RNA (**mRNA**), ribosomal RNA (**rRNA**), and transfer RNA (**tRNA**).

mRNA is the template for protein synthesis. Proteins are polymers of molecules termed **amino acids**, of which there are approximately 20 types within mammalian cells. The sequence in which the amino acids are polymerized together determines the characteristics of the protein; the sequence of the four molecules in the mRNA directs the sequence of amino acids in the protein encoded by the gene, which is described in more detail below. rRNA is a structural molecule that combines with specialized proteins to form the **ribosome**, a globular structure fundamentally important to protein synthesis (Figure 7.5). tRNA is a small, highly structured molecule that recognizes and transports the specific amino acids needed for the growing protein chain. Together, the RNAs comprise an efficient, assembly line factory that produces proteins. The process is remarkably complicated and well controlled.

It begins with signals to synthesize RNA. A seemingly boundless repertoire of signals are involved in turning off and turning on RNA synthesis from genes. Hormones, growth factors, invasion by bacteria or viruses, starvation, gluttony, and sunshine are examples of stimuli for gene expression. Each stimulus excites an intracellular signal transduction pathway that, principally through a series of phosphorylations and dephosphorylations (the domino effect described in Chapter 2), in turn mobilize **transcription factors** to turn on or off gene expression in the nucleus. Some transcription factors are proteins that bind specifically to an area of the chromosome and expose the gene to the RNA polymerases. In addition to RNA polymerases, transcription involves a large complex of enzymes and other co-factors that bind to a specific region of the gene, termed the **promoter**, and systematically manufacture the RNAs, which then leave the nucleus for the cytoplasm. Multiple RNAs can be transcribed from the same gene at the same time. rRNAs combine with ribosomal proteins in the cytoplasm. The tRNAs coil up on themselves and remain free spirits to search the cytoplasm for their specific amino acid.

Triplet codon: the sequence of three bases in mRNA that define an amino acid in the encoded protein; the mRNA triplet codon aligns with the specific transfer RNA (tRNA) bound to the corresponding amino acid.

mRNA slides between the two subunits of the ribosomes (Figure 7.5), exposing its sequence in groups of three molecules, the **triplet codon**, which defines an individual amino acid. For example, the sequence GGU specifies the amino acid glycine, the sequence AUU specifies isoleucine, the sequence GUU specifies valine, and GAA specifies glutamic acid. Together, the sequence GGUAUUGUUGAA specifies the amino acid chain glycine-isoleucine-valine-glutamic acid, the first four amino acids in the hormone insulin.

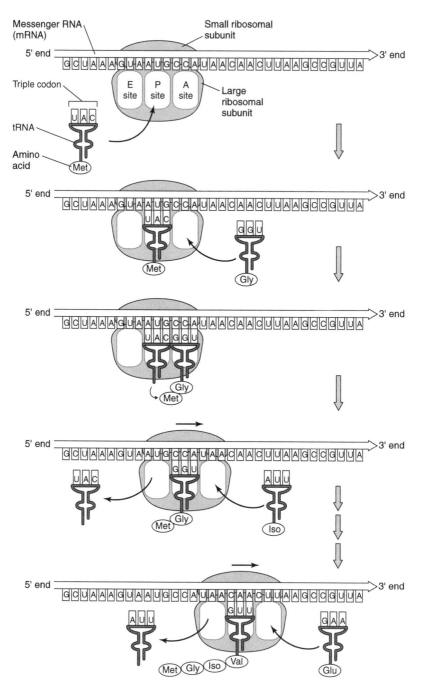

Figure 7.5 Schematic of translation.

The process of reading the mRNA sequence and synthesizing the corresponding protein is termed **translation**.

The triplet attracts its amino acid by binding with a specific tRNA that, in turn, has found its amino acid. The amino acid is brought into the ribosomal-mRNA complex and positioned as the next amino acid in the chain (Figure 7.5). It is chemically bonded to the growing chain like the next piece in a jigsaw puzzle. It all happens very fast. Proteins are synthesized at the rate of 20 amino acids per second. Thus, one molecule of insulin can be synthesized in approximately 3 seconds. Each mRNA can simultaneously direct the synthesis of multiple growing protein strands, and each gene can transcribe multiple mRNAs simultaneously. Thus, millions of copies of proteins can be produced from one gene sequence in a matter of minutes, and production can be halted just as fast. The speed, precision, and responsiveness of gene expression is an awe-inspiring cellular asset.

In no cell is it more finely tuned than in the egg. As described in Chapters 3 and 4, unlike dividing somatic cells that synthesize and degrade proteins, as directed by either cell cycle machinery or specific cellular functions, the egg synthesizes large stockpiles that must be deployed as needed. Ribosomes and ribosomal precursors are stockpiled in massive quantities by the egg. Large stores of mRNAs, termed maternal messages in eggs, such as cMos, which is essential for metaphase II arrest (sidebar, Chapter 3), are also stockpiled. Other cellular components essential for rapid and accurate early cleavages are stored as proteins, such as the DNA polymerase detected in relatively high concentration in metaphase II-arrested mouse eggs. The systematic deployment of these stockpiles for the multiple cycles of DNA replication and cleavage is a property of eggs that remains poorly understood, especially with respect to human eggs.

Transcription and Translation in Frog Eggs

The nuclear transfer experiments conducted in the 1950s led to an appreciation of the frog egg as an experimental system with which to explore controls on gene expression. Frog eggs can be recovered without damage to the frog (they can be hugged out by a belly squeeze, mimicking the bull frog), and thus represent an unlimited supply of eggs for study.

Injection of specific mRNAs into frog eggs, such as the mRNA for human hemoglobin, resulted in the synthesis of hemoglobin proteins in the frog egg cytoplasm (Figure 7.6). In addition, injection of genes in the form of DNA fragments were transcribed into messages and then translated into proteins. Several groups of investigators studied the transcription and translation of genes and mRNAs injected into frog eggs. The results of their studies led to valuable new insights into the controls on the processes of transcription and translation.

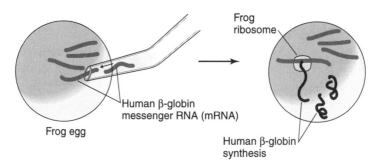

Figure 7.6 Injection of human mRNAs into frog eggs.

Polyadenylation: the process of adding a series (sometimes hundreds) of adenine nucleotides to the end of an mRNA molecule to signal it's binding to ribosomes and translation into a protein.

One important control is the need to add a string of As at the beginning of the mRNA to signal that it is prepared for translation. This process is termed **polyadenylation**, and is now recognized as an important control of gene expression. mRNAs that are not polyadenylated may thus be stockpiled and not translated into protein. Therefore, one way to recruit specific maternal messages from the egg's stockpile is to bring about polyadenylation. This is now recognized as an important egg control mechanism for both egg activation and early cleavages.

Protein stockpiles may also have specific control mechanisms. After translation, many proteins require additional modifications before they become functional. Such modifications include splitting off a small segment that inhibits activity, or the need to add side chains, such as fatty acids or carbohydrates, to allow the protein to be transported to its point of action. For example, insulin is synthesized as a long precursor protein that is cleaved into two polymers that then interact with each other in a specific fashion to bring about folding of the molecule to fit precisely into the insulin receptor. Precursor proteins are not active, and thus can be held in reserve until needed.

These considerations highlight the many possible mechanisms available to an egg both to stockpile cellular components and to control their deployment. This is in sharp contrast to somatic cells whose gene expression, both on and off, is more tightly controlled by environmental stimuli. Size differences alone between the germinal vesicle and somatic nuclei suggest different mechanisms of control of gene expression. Somatic cell chromatin must be opened up by transcription factors to allow transcription to occur, whereas the huge scaffolding available to egg chromatin in the germinal vesicle may obviate the need for transcription factors. Transferring a nucleus from a somatic cell into an egg is, therefore, not simply transferring a bag of genes. The somatic cell genes are under an entirely different set of regulations that

Nuclear remodeling: changing a somatic cell nucleus into the large nucleus of a zygote or cleaving blastomere.

must be set aside to allow the egg cytoplasm to take over control of DNA synthesis and cell cleavage. This process has been termed **nuclear remodeling**.

Reprogramming a Nucleus

Genes can be broadly divided into three groups: housekeeping genes, cell cycle genes, and cell function genes.

Housekeeping genes encode proteins that contribute to routine cell functions, such as respiration, protein synthesis, protein degradation, and DNA repair. Many of these genes are highly conserved between species, some even between plants and animals. Some are expressed at all times, termed **constitutive** expression, but others are expressed only at specific times, such as heat shock proteins, which appear with fever. It is presumed that, for the most part, housekeeping genes may not need to be reprogrammed to allow the nucleus from a differentiated cell to direct embryonic development.

Cell cycle genes were discussed in Chapter 2 (Table 2.1). Although many aspects of blastomere cell cycles are similar to somatic cells, there are distinct differences, especially related to the foreshortened G1 phase. So, although cyclin synthesis and degradation appear to be highly conserved between species, and between embryonic and somatic cells, the details of cell cycle regulation in a somatic cell need to be reprogrammed for successful and sustained blastomere cleavage.

Cell function genes are the essence of a differentiated cell. Expression of specific sets of genes define cell types. Some cells express genes not expressed in any other types of cells. For example, the proteins that comprise the lens of the eye are not expressed by other cells. The zona pellucida proteins that provide the protective coating for eggs are not expressed by other cells. Hemoglobin, the protein that carries oxygen to all cells of the body, is expressed specifically in red blood cells. Cell surface receptors for hormones, such as follicle-stimulating hormone (FSH, Chapter 3), are expressed by only a few cells, principally those in the ovary and the testis. The immune system is redundant with examples of cell surface receptors that enable a specific immune responsive cell to mount an attack against an infectious agent.

These examples highlight the nearly incomprehensible spectrum of cellular behavior encoded by the human genome. In contrast, do the egg and cleaving blastomeres express fewer genes and in this way avoid such highly differentiated cell functions? The egg itself is a highly differentiated cell, as is the sperm. So is it the combination of two new genomes that avoids differentiation into committed cell types for the first several cleavage divisions? These are the questions that have plagued developmental biologists for decades, and their answers will provide valuable information about mechanisms necessary to bring about nuclear reprogramming.

What is clear, with respect to nuclear reprogramming, is that the highly specialized cell functions need to be turned off, and the nuclear chromatin needs to be prepared to undergo S phase nearly as soon as the nucleus is injected into the egg. This is undoubtedly a complicated requirement.

As discussed previously, HeLa cell nuclei were injected into frog eggs in 1974 in the Gurdon laboratory. Human mRNA and proteins were detectable for several days following the nuclear transfer. This landmark experiment demonstrated several fundamentally important findings:

1. Frog eggs do not contain foreign DNA destructive enzyme systems, as do many bacteria; this was a somewhat surprising finding given the importance of maintaining the species specific integrity of the genome. The lack of an inherent mechanism for destroying foreign DNA seems especially curious for an animal whose eggs are fertilized among the many foreign species present in pond water. This observation is also true for eggs from other species, a finding supported by Ben Brackett's report that rabbit sperm could carry radioactive viral DNA (SV40) into rabbit eggs at fertilization. These two observations are consistent with the lack of detectable DNA-degrading enzymes (**deoxyribonucleases**) in mouse eggs. Zygotes also may not have a mechanism for destroying foreign DNA given the ease with which transgenic mice are produced (sidebar, Chapter 5).

2. Frog transcription factors and RNA polymerases can transcribe at least a subset of human genes, supporting the highly conserved nature of at least some gene expression controls across several animal phyla. Numerous experiments injecting human nuclei into frog eggs followed these early experiments, with multiple reports of expression of specific human genes. Studies to specifically reveal human genes that cannot be expressed in frog eggs are lacking.

3. Frog ribosomes, translation factors, and tRNAs can recognize and synthesize human proteins from the human mRNAs, providing further support for the highly conserved nature of mechanisms of protein synthesis.

In summary, although the frog egg experiments have yielded enormous information about egg controls on gene expression, the transfer of the HeLa cell nuclei did not lead to cleavage. Determining whether or not this was due to an incompatibility between frogs and humans, or whether mammalian somatic cell nuclei were fundamentally different from frog somatic cell nuclei, would require the development of nuclear transfer techniques for work with mammalian eggs, which will be discussed in Chapter 10.

Additional Reading

Davidson, E. H. (1986). *Gene Activity in Early Development*. London: Academic Press.

PART III

Embryonic Stem Cells

CHAPTER 8

The Blastocyst and Inner Cell Mass Cells

Indeed, each blastomere has the potential to form any cell of the body.

James Thomson

OVERVIEW

As described in Chapter 6, the geometry of the cleaving blastomeres changes with each cell division. Four-Cell blastomeres are all the same size, but Six-Cell blastomeres are two different sizes: four are smaller (Eight-Cell size) and two are larger (Four-Cell size) (Figure 8.1). Thus, only at the four-, eight- and 16-cell stages are all the blastomeres the same size. Because of the spatial constraints within the zona pellucida, the blastomeres rearrange themselves with each cleavage event. Each of the eight-cell stage (early morula) blastomeres has contact with other blastomeres and the outside of the cell cluster. However, by the 16-cell stage, one or two of the blastomeres is completely trapped inside the cell mass with and has no contact with the outside world (Figure 8.1).

The enclosure of a cell, or a couple of cells, inside the morula signals the first embryonic commitment event: The cells on the outside become committed to giving rise to the placenta; those trapped inside will give rise to the embryo. The outside cells are termed **trophoblast**, the foundation of the placenta; the inside cells are termed **inner cell mass (ICM)** cells. Under some circumstances, ICM cells can contribute to the trophoblast layer, but trophoblast cells are not thought to contribute to ICM. Best studied

ICM: *inner cell mass*, the cluster of cells at one end of the blastocyst that gives rise to stem cells.

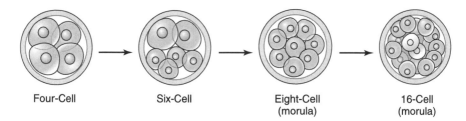

| Four-Cell | Six-Cell | Eight-Cell (morula) | 16-Cell (morula) |

Figure 8.1   Six-cell stage blastomeres are unequal in size.

in the mouse, there is evidence to embrace similar commitment events in the rabbit, cow, sheep, and presumably human. The precise cellular mechanisms that dictate this first commitment event remain elusive. The importance of fully understanding this critical developmental event cannot be overstated.

The decision to become either an ICM or trophoblast cell is characteristic of the type of commitment decisions that occur throughout embryonic development. One of two paths is chosen, and once chosen, as is the case with trophoblast, back differentiation does not seem to occur. Only those cells that have not taken the next step of differentiation remain available for the full spectrum of future differentiation events. These are also the basic properties of stem cells. It is speculated that specific sets of genes are turned off and new sets turned on at each differentiation event, but evidence for this is incomplete. This is an active area of investigation, with new information reported regularly. What is known is that the ICM cells and the trophoblast cells develop distinct characteristics with respect to cell cycle times and gene expression. The embryonic genome is clearly expressed by this stage, but some important egg functions persist as well.

Curiously, laboratory conditions for culturing individual blastomeres have not been developed. Alone, the blastomeres do not continue to cleave. It is possible that, because of their large size, they need physical as well as nutrient support for their cell cycles. In addition, progression through their cell cycles may require stimulation by factors they secrete themselves. Termed autocrine stimulation, described in Chapter 6, this is a well known mechanism for several cell functions. Such factors, usually proteins, are synthesized in the cytoplasm and secreted through the plasma membrane. Once outside the cell, they bind to specific receptors on the surface of the plasma membrane that are primed to interact with signaling molecules inside the cell. The signaling molecules bring about transitions in the cell cycle (described in Chapter 2). Although speculative, there is evidence that such factors are produced and are important to blastomere cleavage. It is possible that one isolated blastomere may not synthesize factors in sufficient concentration to activate receptors and promote progression through the cell cycle. Until this problem is solved, the fate of blastomeres that do not experience either an inside or outside position in the morula will not be known. In addition to providing fundamentally important information about early developmental commitment events, developing the technology to culture individual blastomeres could greatly enhance the efficiency of deriving stem cells from activated eggs, as will be discussed in Chapter 10.

Equivalence Groups

Studies reported in 1979 and 1980 of the simple microscopic flatworm, *Caenorhabditis elegans*, revealed that from a group of six cells that were precursors to its egg-laying structure, only three cells gave rise to it. The other cells then went on to contribute to primitive skin. However, ablation of the three cells initially committed to the egg-laying structure caused the remaining three to abort their skin destination and give rise to the egg-laying structure instead. This showed that the six cells were multipotential and could compensate for each other if necessary, thus giving rise to the concept of an **equivalence group** of cells. This is an attractive theory that carries with it the caveat that equivalence groups have a primary (default) fate and a secondary fate that is assumed only if not needed for the default. Once the secondary fate is assumed, however, the primary fate cannot be reassumed, thus establishing a time frame within which decisions must be made. In this way, the direction of cell commitment events is also established. This concept seems fundamentally important and redundant throughout embryonic development.

The Blastocyst

As described in Chapter 6, the transition from cleaving blastomeres to the ball of cells termed the morula happens at approximately the same time that the embryo leaves the oviduct and enters the uterus. It is as if the oviduct provides an environment protected from outside influences, such as serum growth factors. In this way, the oviductal

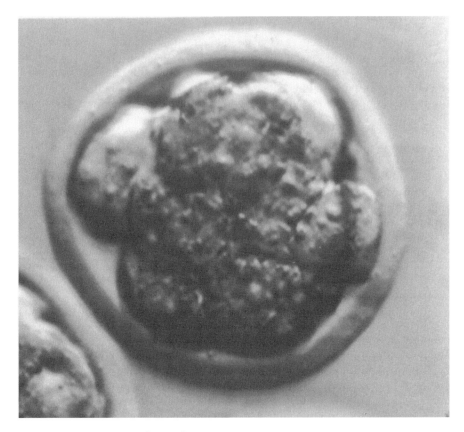

Figure 8.2a Compacted morula.

environment allows the cleaving egg to accomplish several replications of DNA, finally achieving a cell size more in keeping with somatic cells. As discussed in Chapter 2, dividing somatic cells must grow in size for each cell division. Factors that stimulate specific cell pathways appear necessary for cell growth, as described in Chapter 6. Since continued cell division is essential for continued development, the embryo must leave the protected environment of the oviduct and enter the growth factor-rich environment of the uterus.

Several things happen once the cells become committed due to their position in the morula. The outer cells develop tight junctions with each other, termed **compaction** (Figure 8.2a). Compaction has been most thoroughly studied in the mouse because it occurs by the 12- to 16-cell stage. Compaction does not occur in other species until somewhat later stages, the 16- to 32-cell stage in the cow, goat, and possibly human. The individual blastomeres flatten against each other to form a cohesive cluster that loses the appearance of a group of individual cells. The expression

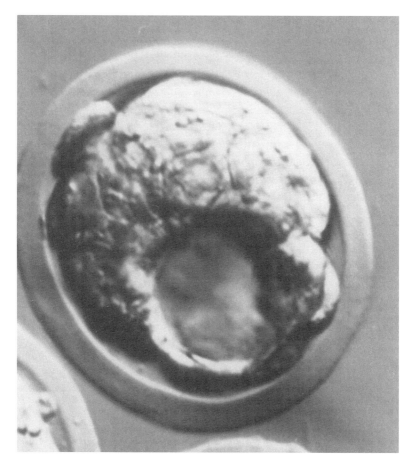

Figure 8.2b Early cavitating mouse blastocyst.

of a specific membrane ion channel enables the outer cells to actually pump water and nutrients in the direction of their interior surfaces. In this way, the cells become polarized. Their physical shape changes; they become elongated and take on the appearance of flattened epithelial cells.

As the cells pump water and ions inward, the fluid collects together on one side of the interior (Figure 8.2b) giving rise to the **blastocyst**. The fluid filled chamber is the **blastocoel** cavity. It is thought to help provide necessary nutrients for the expanding ICM. The trophoblast cells opposite the ICM (termed the **abembryonic** end of the blastocyst) cease to divide and begin to accumulate lipid deposits. They also undergo **endoreduplication**, a

Blastocyst: comprises the ICM and an outer layer of trophoblast cells sealed together to maintain the fluid-filled blastocoel.

process by which the DNA is replicated many times without cell division. The value of this is unclear, but may be related to the need to maintain a sealed cavity since the process of mitosis, of necessity, must break the gap junctions that comprise the seal between the trophoblast cells. Endoreduplication, therefore, provides a mechanism for increasing the copy number of genes, and hence gene expression, without the process of mitosis, which would break the seal.

It is important to note that expression of some of the genes involved in the formation of the blastocoel such as the ion channel that is responsible for the transport of water, is not dependent on the number of cell divisions that have occurred, but on the length of time following activation of the egg. Some mouse blastomeres that undergo cell cycle arrest are nonetheless capable of pumping water into vacuoles within their own cytoplasm approximately 94 hours after egg activation, the time at which the blastocoel cavity first appears in developing mouse embryos. This is another example of the cascade of cellular reactions that results from the systematic deployment of components stored in the egg's cytoplasm. This observation makes it all the more curious that ICM cells do not also express the ion channel at the same time it is expressed by trophoblast cells. One possible explanation for the difference is the existence of both positive and negative controls on deployment of egg stores, some being deployed by "outer" cells, and others by "inner" cells. This line of reasoning indicates that maternal cytoplasmic stores are in some way responsible for activation and control of the embryonic genome. Nuclear transfer studies described in Chapter 10 will explore this possibility in more detail.

Primitive Endoderm

When the ICM of the mouse is comprised of approximately 50 cells, and the entire blastocyst contains approximately 150 cells, a second differentiation event occurs. The layer of cells that line the blastocoel side of the ICM take on a new, flattened appearance and form the primitive endoderm, the layer of cells that will contribute to the **extraembryonic membranes**, including the **amnion**, the fluid-filled sac that houses the developing fetus. This second differentiation event occurs at approximately the same time the mouse blastocyst sheds its zona pellucida, a process called **hatching**. Once outside the zona pellucida, the blastocyst can attach itself to the lining of the uterus. The interaction between the blastocyst and the uterine lining is termed **nidation** and heralds the process of **implantation**. This, therefore, marks the end of the **preimplantation** period of development.

Implantation: the process of trophoblast attachment and outgrowth into the lining of the uterus.

Implantation

Signals between the cells of the uterus and the trophoblast cells of the blastocyst encourage the trophoblast cells to physically attach to and begin to invade the uterine epithelium. This is the beginning of the placenta.

There are many similarities between trophoblast invasion of the uterine epithelium and a cancerous tumor. The principal difference is that growth stops once the appropriate size of the placenta is reached, whereas a malignant tumor continues to grow.

Once implantation has occurred, the embryo doubles in size daily and organogenesis begins, as will be discussed in Chapter 9.

Blastocyst Development In Vitro

It is important to remember that no existing laboratory conditions support fully normal development of pre-implantation embryos of any species. First studied in the mouse, rabbit, and hamster, several characteristics of in vitro-developed embryos have become clear.

First, blastocysts developed in culture contain markedly fewer cells than blastocysts developed in utero (Figure 8.3). An analysis of the cell cycles of cultured

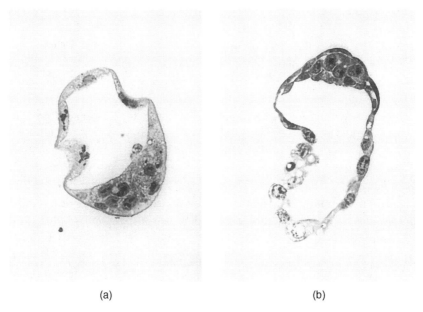

(a) (b)

Figure 8.3 Comparison of mouse blastocysts developed in culture (a) and in the mouse (b).

mouse embryos reveals that the zygote stage is approximately the same length, but the two-cell stage is a few hours longer, as is the four-cell stage. The net result of longer cell cycle times during the first few cleavages is that blastocysts form at the species-appropriate time after egg stimulation, but they contain only one half to one third the number of cells. For example, hatching blastocysts developed normally in the mouse have approximately 150 cells 120 hours after fertilization. In contrast, mouse blastocysts developed in the laboratory have only one half to one third that number of cells. Comparisons of DNA polymerase activities in mouse embryos developing in culture and developing in the oviduct or uterus reveal similar levels per embryo, so a lack of DNA-synthesizing enzymes is not the cause of the increased cell cycle times and the decreased cell numbers in the cultured embryos.

Second, the percentage of cells that comprise the ICM remain the same in cultured blastocysts, approximately 30% for the mouse. Since there are fewer total cells, the net result is that cultured blastocysts contain smaller ICMs than embryos developed in utero. Some blastocysts developed in culture contain no ICM cells at all; they are just spheres of trophoblast cells. It is possible that such spheres are analogous to the "empty sacs" observed in some pregnant women whose pregnancies never develop past early implantation. Historically, obstetricians have referred to such early miscarriages as "blighted ovum" in reference to trophoblast cells with no attendant embryonic cells. Given the potential for eggs to activate and initiate cell cleavage under a variety of conditions, there could be multiple reasons for such defective development.

Third, blastocysts that develop in vitro rarely, if ever, develop a layer of endoderm on the inner aspect of the ICM (Figure 8.3). This could simply be because the numbers of ICM cells never reach the critical mass necessary to stimulate the second differentiation event. The fact that primitive endoderm formation is retarded may actually be an advantage for the derivation of lines of embryonic stem cells from ICMs with the potential to differentiate into endodermal structures. Alternatively, it is possible that successful lines of stem cells are only derived from ICMs that do develop primitive endoderm in culture, but not in sufficient cell numbers to be identified as endoderm.

Fourth, it is important to emphasize that a higher percentage of fertilized, or otherwise activated, eggs will develop to blastocysts in culture than in the uterus. Several studies have demonstrated that most of the fertilized eggs in a hormone-treated mouse will develop to blastocysts in culture, but many will fail at the early morula stage in the uterus. It is as if uterine signals stimulate programmed cell death of defective embryos. This is an important area of research that has received little attention, but could provide valuable clues about normal developmental signaling between the uterus and pre-implantation embryos.

Embryonic Stem Cells

In a landmark report more than 20 years ago, Martin Evans and Martin Kaufman, at the University of Cambridge, England, reported the isolation of pluripotent embryonic stem cells from mouse blastocysts. They reasoned that prior attempts to obtain lines of continuously growing, nondifferentiated cells had failed because insufficient numbers of ICM cells were obtained and because culture conditions had been more supportive of cell differentiation than of cell division. They took advantage of a fascinating aspect of embryonic development that exists in a number of mammalian species, that of **delayed implantation.**

Many mammals, such as rodents and migratory marine mammals, ovulate within a few hours of giving birth, presumably in response to the dramatic drop in pregnancy hormones. Those eggs may be fertilized if she mates with a fertile male, but as long as the mother is producing milk, the fertilized eggs do not implant in the uterus. This is termed **lactational delay**, or **embryonic diapause.** The fertilized eggs exist in the uterus as free-floating blastocysts, many of which hatch from the zona pellucida and gradually increase in cell number. Overall, their metabolism slows down. Delayed mouse blastocysts have decreased DNA and RNA polymerase activities, reminiscent of cells that have entered G0. Most of the increase in cell number occurs gradually in the ICM, thus producing blastocysts with especially robust ICM numbers, including some differentiation of primitive endoderm. Under normal circumstances, when the mother weans her pups, the decrease in the hormones responsible for milk production leads to an outpouring of estrogen from the ovary, which in turn stimulates implantation of the delayed blastocyst and the resumption of development. This is an impressive example of the powerful adaptive mechanisms that have evolved to ensure survival of many species.

Delayed implantation can be mimicked in laboratory mice by the administration of hormones to a pregnant mouse. The blastocysts will exist in a free-floating state for up to 3 weeks under these conditions. Evans and Kaufman recovered delayed implanting blastocysts and placed them in culture. Within days, the blastocysts had attached to the culture dish and the ICM cells had formed large egg cylinder–like structures comprised of small round cells surrounded by endodermal cells. The egg cylinders were dispersed into individual cells and the cell suspensions added to culture dishes that already contained a layer of fibroblasts as a feeder layer (discussed in Chapter 6). Actively proliferating colonies of cells were apparent from an early stage and mass cultures of the ICM cells could be produced.

These investigators had previously studied the cell lines produced by culture of ovarian dermoid cysts that occur spontaneously in some strains of mice, as discussed in Chapter 4. Although not all such cell lines were derived from malignant tumors, they are generally referred to as **embryonal carcinoma (EC)** cells and will be discussed in Chapter 9. EC cells proved to be very useful experimental systems in many respects, but two important features, their lack of uniform chromosome number and their lack of a Y chromosome, limited their usefulness for some types of experiments. It is for this reason that Martin and Kaufman were keen to derive a cell line with a normal chromosome number from a male embryo. One of the first studies they did with the cell lines derived from the delayed implanting blastocysts was to measure chromosome number, which they found to be normal. They termed these new cells **EK cells** to distinguish them from EC cells, but that acronym was quickly replaced by **ES cells** for **embryonic stem cells.**

The same year, another developmental biologist, Gail Martin, also successfully cultured mouse ICM cells to derive a pluripotent stem cell line. She chose culture medium conditioned by EC cells instead of a monolayer of cultured cells. Dr. Martin had previously studied EC cells and had contributed to the understanding that EC cells were pluripotent by creating offspring that were **chimeric**; this is described in the section to follow.

Importantly, the ease with which ES cells can be derived from the ICM of mouse blastocysts is highly strain specific. The mouse strain chosen by Evans and Kaufmann and Martin, 129J, has proven to be the most successful for deriving ES cells. To date, the reason for this is not known, nor is it known if it is related to the high rates of spontaneous activation of eggs in this mouse strain that lead to the dermoid cysts and teratomas discussed in Chapter 4. Understanding the differences in ease of ES cell derivation between mouse strains may be critical to understanding how to derive therapeutic stem cells through nuclear transfer.

Chimeras

An interesting experimental technique reported independently in 1961 by Andrei Tarkowski, working in Poland, and in 1962 by Beatrice Mintz, working in Philadelphia, involved combining the blastomeres from two cleaving mouse eggs into one mass before compaction (Figure 8.4a). The resulting morula, a **chimera**, developed into a blastocyst with cells from each original egg in both the trophoblast and the inner cell mass. This elegantly simple experimental approach demonstrated not only the pluripotential nature of early cleaving blastomeres, but also established the possibility of obtaining mice with tissues made up of two genetically independent cell types. This has proven to be a powerful experimental technique with which to explore the contribution of individual cells to overall development of the mouse.

An early modification of the technique was reported by Richard Gardner in 1968 in which chimeric offspring were obtained following injection of blastomeres into the blastocoel cavity of a mouse embryo just before implantation (Figure 8.4b). Viable offspring were obtained whose tissues were chimeric, and included

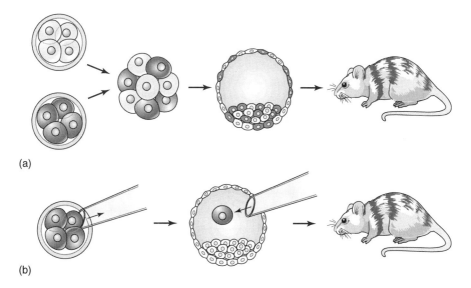

(a)

(b)

Figure 8.4 Construction of mouse chimeras.

both the cells arising from the fertilized egg and those injected into the blastocoel cavity. This work not only provided further proof of the pluripotency of early embryonic cells, but also evidence of the accepting nature of ICM cells to the incorporation of "foreign" cells into fetuses. This remarkable plasticity of embryonic development had not been previously appreciated, and heralded the new era of manipulating individual mouse genes to more fully understand their function (see sidebar).

In summary, the first cell commitment event occurs when one blastomere becomes completely enclosed within the morula. The molecular mechanisms controlling this commitment are poorly understood. The ICM cells retain the potential to develop into all the cells and tissues of the developing organism. Developmental pathways are remarkably plastic considering the need for accuracy in differentiation events. This apparent conundrum may be resolved by the existence of equivalence groups, which are groups of cells with a primary destiny and at least one secondary destiny, if not needed for the primary destiny. Thus, tissues have more than one opportunity to gain the needed number of committed cells. The existence of equivalence groups may also provide a mechanism for the incorporation of foreign blastomeres into a wide variety of tissues following injection into the blastocoel cavity. This not only proves the uncommitted status of blastomeres, but also the lack of rejection of foreign embryonic cells introduced during development. It is this very plasticity that predicts the therapeutic value of embryonic stem cells.

KNOCKOUT MICE

The ability to modify specific genes in order to determine their function and importance to the organism, as illustrated by the mutant yeast experiments described in Chapter 2, has been widely appreciated by mammalian biologists for decades. One approach to this was the development of a technique termed *gene targeting by homologous recombination*.

The concept is elegantly simple: fragments of DNA are taken up by cells in culture and occasionally recombine with their homologous regions in the cell's own DNA. Although rare, such recombination is not as rare as a spontaneous mutation in a single gene. The DNA fragments can be engineered to contain mutations in specific genes, as well as DNA sequences that encode products that can be detected to determine the success of the recombination event. For example, a bacterial gene for an enzyme, beta-galactosidase, can be inserted within a mammalian gene, such as the one that encodes the enzyme hypoxanthine-guanine phosphoribosyl transferase (HPRT), which is important in preventing Lesch-Nyan syndrome. This type of DNA construct would result in a gene that cannot express HPRT, but can express the bacterial beta-galactosidase that is detectable by a simple staining procedure. Since the ends of the construct would be homologous to the normal HPRT gene, it could replace the normal gene during DNA replication by homologous recombination. Hence, the term gene-targeted homologous recombination. T cells in culture, including mouse ES cells, actually become genetically modified in this manner, as was demonstrated by several groups of scientists throughout the 1980s.

The next step was to determine if mouse ES cells, genetically modified by gene-targeted homologous recombination, could contribute to the formation of chimeric offspring. Three groups of scientists reported their results in 1989.

The first report appeared in early 1989 from the laboratory of David Melton in Edinburgh, Scotland [1]. He and his colleagues studied the mouse model for Lesch-Nyan syndrome, animals genetically deficient in the HPRT gene. They derived ES stem cells from the blastocysts of HPRT-gene deficient mice and corrected the gene defect by adding a DNA construct containing a normal gene. They selected those ES cells in culture that had incorporated the new DNA construct into their genome, and injected the ES-HPRT-normal cells into the blastocoel cavities of embryos that were HPRT deficient. The injected blastocysts were returned to pregnant mothers for gestation and the offspring tested for the presence of the normal HPRT gene. Those offspring that contained the normal gene were then mated, and their offspring also tested for the normal HPRT gene. Those animals that could pass the normal gene to offspring obviously had at least some germ cells that contained the normal gene. Such animals are termed *founders*. This report laid the foundation for correcting genetic defects with ES cell technology.

Two additional reports appeared in the fall of 1989. One was from the laboratory of Bruce Spiegelman, Harvard Medical School, in collaboration with Michael Greenberg, also of Harvard Medical School and Virginia Papaioannou, Tufts Medical School. This team sought to determine if genes that were not expressed in mouse ES cells could be targeted by homologous recombination. They chose three genes (*cfos*, adipsin, and aP2), whose protein products were not detected in mouse ES cells, for gene targeting by homologous recombination. They mutated the three genes in DNA constructs so as to render them nonfunctional. They then selected ES cells that had undergone homologous recombination with the nonfunctional genes for injection into normal blastocysts. Offspring that contained the mutated genes were then mated with the hope of producing offspring that were negative for the genes. Their experiments were successful [2] and they were able to derive founders with each of the three genes "knocked out" in their germ cells. These experiments demonstrated the power of this approach in studies of function and importance of specific genes.

The third report appeared in November 1989 from the laboratories of Stephen Goff and Elizabeth Robertson, Columbia University, New York [3]. These investigators generated mice with mutations in a gene known to play an important role in many cancers, c-*abl*. Although relatively rare, this team was able to obtain several founder animals with specific mutations in various regions of the c-*abl* gene. Most offspring homozygous for the mutated gene did not survive development, thus illustrating the importance of c-*abl* in normal development. These experiments demonstrated the ability to knock out various regions of genes to study gene function.

These elegant experiments demonstrated the power of knock out mice to help determine the specific roles of specific genes, as well as paving the way for applying the technique of gene-targeted homologous recombination to correct genetic defects in ES cells derived from individuals with genetic diseases, such as diabetes.

References

1. Thompson, S., A. R. Clarke, A. M. Pow, M. L. Hooper, D. W. Melton. 1989. Germ line transmission and expression of a corrected HPRT gene produced by gene targeting in embryonic stem cells. *Cell* 56: 313–321.

2. Johnson, R. S., M. Sheng, M. E. Greenberg, R. D. Kolodner, V. E. Papaioannou, B. M. Spiegelman. 1989. Targeting of nonexpressed genes in embryonic stem cells via homologous recombination. *Science* 245: 1234–1236.

3. Schwartzberg, P. L., S. P. Goff, E. J. Robertson. 1989. Germ-line transmission of a c-abl mutation produced by targeted gene disruption in ES cells. *Science* 246: 799–803.

Additional Readings

Johnson, M. H. (1977). *Development in Mammals*. Amsterdam: Elsevier/North Holland.

Kiessling, A. A., Davis, H. W., Williams, C. S., Sauter, R. W., and Harrison, L. W. (1991). Development and DNA Polymerase Activities in Cultured Pre-implantation Mouse Embryos: Comparison with Embryos Developed in Vivo. *J Exper Zool* 258: 34–47.

Wolstenholme, G. E. W. (1965). *Preimplantation Stages of Pregnancy*. Boston: Little, Brown.

CHAPTER 9

Organogenesis

Listen to me, little fetus,
Precious homo incompletus,
As you dream your dreams placental
Don't grow nothing accidental!

Poem by an anonymous father

OVERVIEW

It seems so unlikely that a fertilized egg could give rise to an entirely new individual complete with functioning organs and brain, that the mere notion is preposterous, and yet it happens over and over. Aristotle attempted to understand embryogenesis through the study of fertilized chicken eggs. The developing chicken, like amphibia, has provided some fundamentally important insights into organogenesis because many characteristics of the early stages of embryogenesis are highly conserved across species. Mammals, however, form placentas and thus require their own detailed study.

John Hertig was a pathologist colleague of Ruth Menkin and John Rock (sidebar, Chapter 4). To more fully understand early human development, Hertig systematically sectioned the uteri of women undergoing hysterectomy by John Rock. Dr. Rock explained his interest in early human development to dozens of women in need of hysterectomy for unrelated reasons. The women agreed to have a hysterectomy at timed intervals following ovulation and intercourse. Dr. Hertig then explored the entire uterus for the presence of an early pregnancy, which was then examined. In this way, Hertig and Rock compiled the most complete description of early human development available to this day. Most of their original materials are in storage at the Carnegie Institute. Given the current social concerns with studies of early fertilized human eggs, we are deeply indebted to their work, for without it we would understand even less than we do now.

In 1956, Hertig, Rock, and Adams described a unique series of "...34 human ova within the first 17 days of development." Hertig summarized his many years of examining early developmental stages as follows: 15% of eggs ovulated by normal, fertile women fail to fertilize, and another 10 to 15 percent fertilize, but fail to implant. Of the 70 to 75% of fertilized eggs that implant, only 58% will survive until the second week and 16% of those will be abnormal. Thus, only 42% of eggs exposed to sperm survive until the first missed menstrual period, and of those a number will be aborted during subsequent weeks. These findings are in basic agreement with more recent large epidemiologic studies that indicate that it takes an average of 3 to 4 months for normal, fertile couples to achieve a viable pregnancy.

It is important to note that at no time were the fertilized eggs referred to as "embryos" by Hertig or Rock. As discussed in Chapter 4, the term embryo was not applied to cleaving eggs until after assisted reproduction became the standard of care for infertile couples. Interestingly, the term zygote was broadly applied to early developmental stages, sometimes even after implantation. The term **zygotic genome activation (ZGA)** has been used to refer to the turn-on of gene expression in cleaving blastomeres. Implied in this use of the term is the notion that an embryo is a stage reached after developmental control is transferred entirely from the egg cytoplasm stores to the new genome created as a result of fertilization. The term is also applied to describe the two types of twins that may arise: **fraternal twins (dizygotic)** arising from two fertilized eggs, and **identical twins (monozygotic)** arising from one fertilized egg that may split into two developing embryos at various stages after the first cleavage and before the formation of the primitive streak, as described below.

Dizygotic: twins arising from two fertilized eggs.

The Second Week

As described in Chapter 8, implantation occurs approximately 7 to 10 days after fertilization. The trophoblast cells overlying the ICM attach to the newly enriched layer of cells that line the uterus (**endometrium**) and begin to invade between them (Figure 9.1). The trophoblast secretes human chorionic gonadotropin (hCG), a protein hormone that maintains the corpus luteum, as described in Chapter Three. hCG may also be a growth factor that stimulates development of the placenta. Once the placenta has reached a critical mass of cells, it synthesizes and secretes the steroid hormones, estrogen and progesterone, necessary to maintain the pregnancy.

A layer of primitive endoderm forms on the inner aspect of the ICM, the remaining cells of which are now sometimes termed **primitive ectoderm**. By the end of the second week, another cavity begins to appear between the primitive ectoderm and the overlying trophoblast. The net result is that the ICM takes on the appearance of a pancake, two cell layers thick, attached at the periphery to trophoblast with a small cavity above and the blastocoel cavity (primitive yolk sac) below (Figure 9.1). This stage is the **bilaminar germ disc** or **epiblast**, which consists of on the order of a few hundred embryonic stem cells. Very little is known about gene expression and controls on development at this stage in humans.

The trophoblastic cells develop rapidly during this stage and differentiate several types of cells that will form the complex and protective cell layers of the placenta, including the **chorionic villous membrane**, which interacts with maternal blood circulation to take in nutrients and release fetal metabolic by-products. As will be discussed below, mature placentas may contain a substantial population of embryonic stem cells.

The Third Week

At approximately the time of the first missed menstrual period, the bilaminar germ disc becomes a trilaminar germ disc. **Mesoderm** appears as a population of small, rapidly dividing cells between the primitive endoderm and the embryonic ectoderm. Mesodermal cells are protected from the outside environment. They initially create two bulging regions that have a slight depression between them, termed the **primitive streak**, which extends from one edge of the disc to approximately the middle, where it stops at a depression termed the **primitive node**. The appearance of the primitive streak defines the orientation of the developing embryo.

Primitive streak: the depression in the ectodermal layer of the embryonic disc caused by mesodermal cells.

The embryonic ectoderm is the **dorsal** (back) side. The primitive node is the lowest point of the spinal cord. Directly opposite the primitive streak is the region that will eventually give rise to the head. Hence, after approximately 15 days of

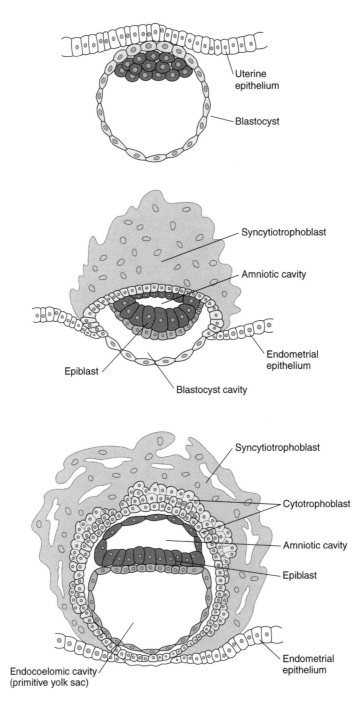

Figure 9.1 Early implantation and formation of the bilaminar germ disc.

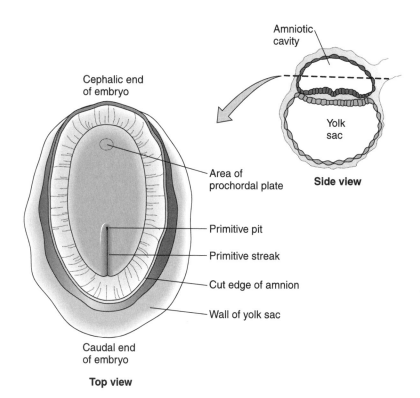

Figure 9.2   The third week.

development, the embryo has a right and left side, as well as head (**cephalic**) and tail (**caudal**) regions and a top (dorsal) and bottom (**ventral**). Things begin to happen even faster. The mesoderm, which in many ways is reminiscent of the original ICM because it has little contact with the outside environment, expands rapidly, causing a thickening and reorganization of the disc. From the primitive pit to the head region, a tubelike area forms within the mesoderm, the **notochord**, which is the precursor to the spine. The embryonic disk takes on a pear-shaped appearance, with most of the expansion in the cephalic region; the primitive pit is thus displaced to the caudal third of the disc instead of in the middle (Figure 9.2).

Continued expansion of the mesodermal layer creates a new midline groove in the cephalic region. The ectodermal cells outgrow on either side of the midline groove to define the **neural plate**, at the apex of which are the **neural crest cells**. This dorsal-reaching growth phase creates ectodermal flaps that reach up and toward each other, eventually touching at the top and fusing at the midline in the center of the embryo. The net result of this growth is to capture a tube of ectodermal cells (the **neural tube**) inside the mesoderm, with a population of neural crest cells located just under the overlying ectoderm (Figure 9.3).

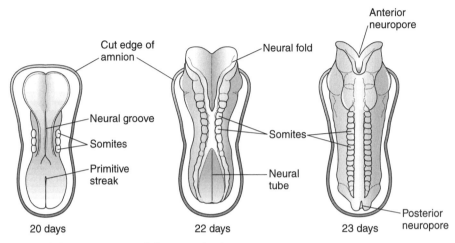

Figure 9.3   Formation of the neural tube.

At the same time the neural tube is forming, bulges of mesoderm are creating **somites**, which appear like two rows of corn kernels on either side of the neural tube. The somites are mesodermal cells that expand in clusters, forming organizing regions for the embryo, which will eventually give rise to skeletal structures such as vertebra and ribs.

The forward growth of the developing embryo has greatly elongated it and positioned the primitive node within the caudal one fifth. As the somites appear, the neural tube closes forward and aft leaving a large and growing opening at the cephalic end, with a much smaller opening at the caudal end. The ventral aspect of the embryo is attached at the periphery to the trophoblastic cells (now the **amnion**) and in the center to what is left of the blastocoel, which has now become the **yolk sac**. Just under the neural tube at the cephalic end, the mesodermal cells have formed a bulge that will give rise to the heart.

Most of the growth of the pregnancy to this stage is in the placenta and extraembryonic membranes. The developing embryo has no heartbeat and only 10 or 12 somites, approximately one third of the total number that will ultimately form to organize the skeletal structures. The bulk of the embryonic cells are stem cells that have undergone few if any differentiation events, although their commitment to ectoderm, mesoderm, and endoderm may be fixed by this stage. The neural crest cells have been studied in several species because they ultimately give rise to many different cell types. Although they are ectodermal in origin, they migrate in unique ways within the mesoderm. There is something about being enclosed entirely within the embryo, not exposed to the outside, which serves to maintain the pluripotency of embryonic cells.

The Fourth Week

It is interesting to note that most medical embryology textbooks refer to the fourth to eighth weeks of development as the **embryonic period**. As discussed previously, the first 3 weeks are sometimes referred to as the zygotic (as in "fertilized egg") period. This terminology serves to emphasize that in the minds of those who study development, the embryo does not exist until at least 21 days after fertilization.

Embryonic stem cells divide approximately once a day, so the developing embryo almost doubles in size daily. This becomes especially obvious during the fourth week. The nutritional requirements of the developing tissues exceed what can be obtained from the cavities surrounding the embryo, so further development depends on successful formation of the primitive heart and the gut (Figure 9.4).

The precursor cells to the blood system come together in the cardiac bulge and form the first heart tube, which begins to pulsate toward the end of the fourth week. It will be several weeks more before the four-chambered heart is fully formed, but the early heart structures exhibit rhythmic contractions that serve to force embryonic blood cells through developing capillaries.

The ventral aspect of the embryo has developed a primitive gut by bringing the lining of the yolk sac (formerly the blastocoel) into closer approximation to the ventral endoderm of the embryo. As early as the end of the fourth week, liver and lung buds are distinguishable as outgrowths of the endoderm. The connection to the yolk sac becomes increasingly restricted as a stalk is formed adjacent to the connecting stalk that will become the umbilical cord.

Most of the somites appear by the end of the fourth week. The rapid growth of the neural tube, the primitive central nervous system, has almost closed the cephalic end of the neural tube to form the beginning of the primitive brain cavities. Formation of

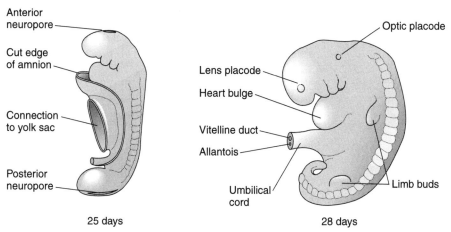

Figure 9.4   The fourth week.

the central nervous system occurs gradually and brain development continues well into the fetal period, extending nearly to the time of delivery. Many embryonic and fetal systems function well in advance of even primitive brain development.

Fifth and Sixth Weeks

The rapid growth of the developing embryo incorporates the endodermal structures completely within the body cavity (Figure 9.5). Lung, liver, stomach, pancreas, and bladder are distinguishable. The growth factors and signals that stimulate their

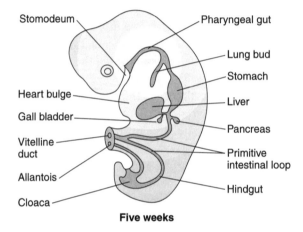

Five weeks

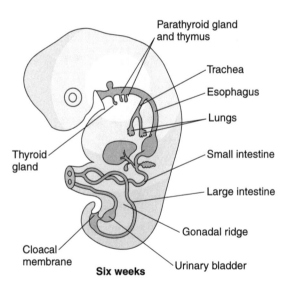

Six weeks

Figure 9.5 The fifth and sixth week.

development are largely unknown. The primitive kidney arises in the mesoderm and is also distinguishable by the sixth week.

By the end of the sixth week, the primordial germ cells have migrated by ameboid movement from what is left of the yolk sac to the gonadal ridge, where their appearance stimulates the development of the testis or the ovary. Expression of Y chromosome genes induce development of the testis and epididymis. In the absence of a Y chromosome, an ovary, oviduct, and uterus develop.

The developing embryo is about half an inch long and weighs approximately half an ounce. Tissue and organ growth is proceeding exponentially, principally because of the expansion of stem cell populations. Some stem cells have become committed to specific tissue types, whereas others remain pluripotent. Few cells are fully differentiated. That process will require most of the 38 weeks of fetal development.

Primordial Germ Cell Migration

Although the development of all organ systems is nothing short of miraculous, preservation of family genetics depends on the successful organization of fetal ovaries and testes. The link between ancestors and descendants depends on the small group of cells, **primordial germ cells**, that make the journey from the yolk sac into the mesodermal region, termed the **gonadal ridge**, at the end of the sixth week (Figure 9.6). The small band of participants and the name of their destination is remarkably reminiscent of the pioneers that settled the American west in the 1800s.

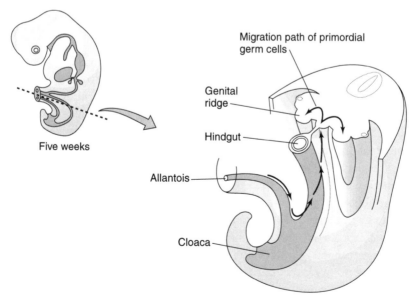

Figure 9.6 Migration of primordial germ cells.

PGC: primordial germ cells, the progenitors of sperm or eggs.

First described in mice (see sidebar), primordial germ cells (PGC) express higher than average levels of an easily detected enzyme activity, alkaline phosphatase, an enzyme that removes phosphate groups (Chapter 2) at pHs higher than neutral (Chapter 5). The role for alkaline phosphatase in PGC physiology is not known, but it has provided a useful PGC marker for numerous tissue studies. In the mouse, approximately 100 PGCs are estimated to reside in the mesoderm on Day 7, approximately 2 days following implantation. Their migration begins shortly thereafter (analogous to approximately 5 weeks following implantation in the human). Remarkably, they proliferate as they migrate, with estimates of tens of thousands eventually populating the gonadal ridge. They migrate over the course of the following week, which represents two thirds of the gestation time of a mouse.

The presence of a Y chromosome initiates formation of the testis and the degeneration of early structures that would otherwise give rise to female reproductive tract organs. In the absence of a Y chromosome, an ovary and associated female structures begin to form. Therefore, by the end of the sixth week, most tissues and organs can be identified and are populated by rapidly expanding stem cell populations, dedicated to the formation of a fully functioning organism. The potential number of misdirections possible during this process seems infinite, but the low number of developmental anomalies present in newborns suggests an equal number of corrective balances must be operational throughout fetal development.

Fetal Tissue Stem Cells

Most of the studies done on harvesting embryonic stem cells from fetal tissues have been conducted in mice. The basic premise has been that embryonic stem cell lines derived from specific tissues will be restricted to the development of those tissues, for example, heart, spinal cord, liver, etc. Experiments have been designed, therefore, to harvest stem cells from tissues of interest and develop the laboratory methods for expanding those cell lines in the laboratory. Since one of the hallmarks of stem cells is that they are precursor cells and, as such, do not perform adult cell functions, another aspect of fetal stem cell research has been to understand the factors that bring about their differentiation and normal adult cell function.

For example, nerve stem cells can be isolated from the neural tube described previously and expanded in culture in the laboratory. Their continual division requires the presence of added growth factors, supporting the concept that they do not synthesize the necessary growth factors themselves. Fibroblast growth factor (FGF) seems to be essential for the continued cell division of a wide variety of neural stem cells.

PRIMORDIAL GERM CELLS AND EMBRYONAL CARCINOMA CELLS

Nearly 50 years ago it was discovered that the strain of mouse that had a relatively high incidence of ovarian dermoid cysts also exhibited spontaneous testicular teratomas in 1% of males. This discovery was made by L. C. Stevens and C. C. Little at the Jackson Laboratories. During the 1960s, this team reported several characteristics of the teratomas that laid the groundwork for current understanding about both embryonic carcinoma (EC) cells and embryonic germ (EG) cell stem cells.

First of all, the incidence of teratomas was markedly increased by transplantation of the gonadal ridge from fetuses into the testes of adult mice. Importantly, the incidence of teratomas from other strains of mice could also be increased by transplantation of the gonadal ridge into adult testes. However, tumor induction was limited to a 2-day interval during development, when many of the germ cells had completed the migration. Once the migration was completed (approximately Day 12 to 13 of gestation), the incidence of tumor formation decreased dramatically. Thus, the ability of germ cells to continue cell division was markedly reduced once the gonadal ridge was fully populated, implying some change in the pluripotency of the cells.

Second, some teratomas could be retransplanted, but others could not. The retransplantable tumors, termed teratocarcinomas, contained EC cells with morphology characteristic of human teratocarcinomas.

Third, both transplantable and nontransplantable tumors contained differentiated tissue elements, such as bone, cartilage, hair, and teeth, similar to elements found in ovarian tumors. Importantly, however, work by L. J. Kleinsmith and G. B. Pierce in 1964 demonstrated that a single cell from a transplantable tumor was found to be sufficient for the growth of a new tumor with the full range of differentiated tissue elements noted in the original tumor.

Taken together, these experiments established that embryonic germ cells could give rise to pluripotential tumors, some of which could become malignant cells with the potential to give rise to multiple cell types. In many respects, these experiments were the first demonstration of embryonic stem cells.

During the 1970s, Davor Solter and his colleagues reported a series of experiments demonstrating that teratocarcinomas also resulted from transplantation of mouse blastocysts to testes or kidneys. Established cell cultures from mouse EC cells were first reported in 1970 by B. W. Kahn and B. Ephrussi. That report was quickly followed by confirmatory work from several laboratories, establishing the pluripotent nature of cells derived both from germ cells and from preimplantation mouse embryos.

By the end of the 1970s, it had been established that both primitive germ cells and mouse blastocysts could give rise to tumors with pluripotential cells that could be induced to differentiate into a variety of cell types. In fact, it was difficult to prevent cell differentiation, just as it is today. Although this work was seminal to present-day stem cell technology, since the origin of most of the cell lines was tumors, the cells themselves were heteroploid, thus limiting their usefulness as stem cells.

As of this writing, stem cells have been isolated from the developing forebrain, the midbrain, and the hindbrain, as well as the spinal cord and the developing cortex of mice and rats. For the most part, multipotent stem cells do not migrate, but appear to maintain a relatively stable population localized to their zones. As some of them begin the process of differentiation, they become more mobile and begin to migrate to selected target sites. This is an example of the many stages of differentiation that some cell types pass through during the course of tissue organization. Each stage may also continue to divide, thus they exist in a multipotent state, more differentiated than their stem cell precursors, but more capable of expanding their populations than the mature, fully functioning cells they are destined to become.

In addition, stem cells have been identified from the developing retina. Characterization of retinal stem cells has led to the further discovery that adult retinas also contain a small population of self-renewing stem cells, a fact not widely appreciated before fetal retinal stem cells were characterized. This is an example of the powerful discoveries that may be made about adult tissues through studies of embryonic stem cells. It is conceivable that by understanding fetal tissue stem cells, it may be possible to devise treatment strategies that would promote proliferation and repopulation of damaged adult tissues through regeneration of their own stem cell reserves.

Another neural stem cell of particular interest is the neural crest cell, which appears at the margin of the developing neural tube, as described previously. The neural crest cell is of particular interest for a number of reasons. It is actually capable of differentiating into several different types of cells, including crossing over from an epithelial cell into mesenchymal cells, which are usually derived from mesoderm cell types. Neural crest cells are highly migratory and have been shown to give rise to pigment-producing cells (**melanocytes**), neurons, glial cells, bone, cartilage, smooth muscle, and blood cells. Studies of this remarkable plasticity have intensified within the past few years in an effort to understand, and take advantage of, the pluripotent nature of this remarkable cell.

In contrast to neural cell precursors that have been isolated from fetal rat and mouse tissues by a number of investigators, stem cells from the earliest discernible islands of blood cell formation (**hemangioblast**) have not been successfully isolated and propagated in culture. This is surprising in light of the fact they were first identified 70 years ago. Hematopoietic stem cells have been isolated, however, by a number of investigators from both fetal and adult animals. Lymphocytes, monocytes, neutrophils, red blood cells, and megakaryocytes (give rise to platelets) all derive from the hematopoietic stem cell (see Figure 1.1). In no other system is the powerful, multipotential nature of stem cells better illustrated.

Organs deriving from endodermal embryonic cells—liver, pancreas, and gut—contain a population of stem cells in adult tissues. Hence, although studies of fetal stem cells from these organs have been reported, the bulk of the work has focused on the isolation of stem cells from adult organs.

Trophoblast is the first cell to differentiate during early embryogenesis, and it subdivides itself into two distinct cell types very soon thereafter. One type remains associated with the ICM (the **polar trophectoderm**) and continues to divide. Opposite the ICM, at the abembryonic pole, the trophoblast cells cease to undergo cell division, but continue to replicate their DNA, becoming **polyploid**, as described in Chapter 6. These distinctions appear to carry over into mature placenta as well. Given the embryonic nature of trophoblast cells, several groups of investigators attempted to isolate stem cells from trophoblastic tissues, particularly from early mouse embryos. Finally, in 1998, a specific combination of growth factors and feeder layer cells supported the isolation of several lines of **trophoblast stem (TS) cells**. If either the growth factors or the feeder cells are removed, the TS cells differentiate into giant cells. In some reports, the feeder layer can be replaced by medium conditioned by the feeder layer (as described in Chapter 6), but these reports are just beginning to appear. The ease with which TS cell lines can be established suggests that this will be a valuable system for studies to understand the factors needed to maintain the undifferentiated state of stem cells.

Polyploid: containing more than the normal number of chromosomes.

In summary, successful embryonic development of tissues and organs requires extraordinary coordination of growth signals, stem cell responses, and appropriate cellular migration. Rich in stem cell populations, embryonic organs and tissues should be a reservoir of pluripotential stem cells, few of which have actually been isolated, cultured, and studied, principally due to a lack of knowledge about the laboratory conditions needed to sustain their growth without cell differentiation.

Additional Reading

Langman, J., and Sadler, T. W. (1988). *Medical Embryology*. Baltimore: Williams and Wilkins.

CHAPTER 10

Mammalian Nuclear Transfer Technology

...except in directions in which we can go too far there is no interest in going at all; and only those who will risk going too far can possibly find out just how far one can go.

T. S. Eliot

OVERVIEW

Once the frog skin cell experiments described in Chapter 7 had been reported, scientists attempted similar experiments with rabbit (Bromhall, 1975) and mouse (Modlinski, 1978) eggs. Advances in microscopy manipulation tools, termed **micromanipulators**, had improved the ability to stabilize and inject mammalian eggs with nuclei directly into the cytoplasm. These investigators transplanted nuclei from early rabbit and mouse embryos, respectively, into fertilized zygotes. In each series of experiments, the zygotes that survived the micromanipulation successfully remodeled the injected nuclei, but only a few cleavage divisions occurred. Developmental arrest was attributed to the fact that the nuclear transfer eggs were tetraploid. They contained both the zygote chromosomes and the chromosomes contributed by the transplanted nuclei.

During the late 1970s and early 1980s, Karl Illmensee worked in the laboratory of Peter Hoppe at the Jackson Laboratories in Bar Harbor, Maine. In 1981, they reported a series of experiments designed to address the concerns about tetraploidy. They transplanted nuclei from mouse blastocyst cells into mouse zygotes whose pronuclei were physically removed by **micropipettes** (Figure 10.1). Thus, the resulting reconstructed zygotes contained only the

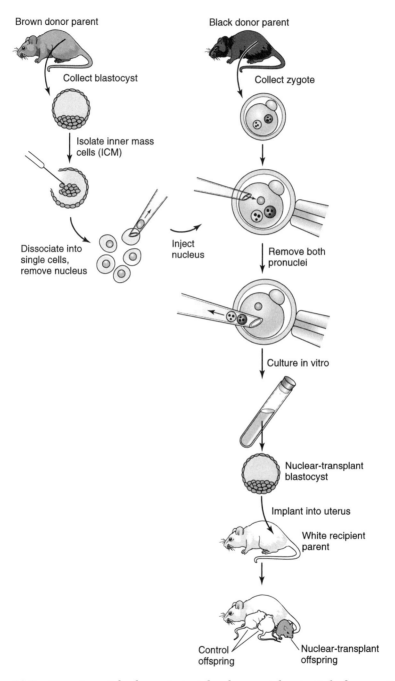

Figure 10.1 Experimental scheme to test developmental potential of mouse inner cell mass (ICM) cell nuclei.

chromosomes from the transplanted nuclei and were diploid, not tetraploid. The results of their experiments, described in the following section, revolutionized thinking about the developmental potential of differentiated nuclei. However, the inability of other investigators to repeat their work cast a cloud on the integrity of their report, which remains today.

The 1981 report by Illmensee and Hoppe addressed two experimental goals: one was to determine if nuclei from trophoblast cells had the same ability to support embryonic development as nuclei from ICM cells, and the second was to determine the potential of the nuclei from cells of the blastocyst to direct the development of a new mouse. They reported that nuclei from trophoblast cells rarely supported development beyond the early cleavage stage: 179 eggs were manipulated, 68 survived both the transfer of trophectoderm nuclei and the removal of the two zygote pronuclei, 34 cleaved at least once, 6 developed to morulae, and 1 developed to a blastocyst (1.5% of surviving eggs). In contrast, nuclei from ICM cells supported a tenfold higher rate of development to blastocysts: 363 eggs were manipulated, 142 survived, 96 cleaved at least once, 25 developed to morulae, and 23 developed to blastocysts (16% of surviving eggs). Moreover, a landmark result was the birth of offspring following transfer of cleaving embryos from transplanted ICM nuclei into the uteri of pregnant mice (Figure 10.1). Thus, in a manner analogous to the frog experiments, mouse ICM cell nuclei could be reprogrammed by zygote stage eggs to direct the development of new individuals. Although the authors did not use the term "clone" in their report, these experiments actually produced the first cloned mice.

In similar experiments conducted the same year, J. Modlinski, working at the Research Council in London, also found that ICM cell nuclei more efficiently supported development to blastocysts following transfer into mouse zygotes than did nuclei from trophoblast cells. Unlike the experiments of Illmensee, however, Modlinski did not remove the pronuclei from the zygotes, illustrating that whether or not the resulting reconstructed embryos were diploid or tetraploid, ICM cell nuclei more efficiently supported cleavage than trophoblast cell nuclei.

Unfortunately, the inability of other research teams to reproduce the work of Illmensee and Hoppe led to accusations that the 1981 report contained falsified data. Following an investigation, both the landmark findings and the scientific integrity of Illmensee were placed in doubt. The accuracy of the report remains unresolved today.

Transplantation of Embryonic Nuclei into Fertilized Eggs

Other research teams were pursuing information about the ability of nuclei from a variety of mammalian cells to support embryonic development. The results from amphibian systems (Chapter 7) suggested that adult cell nuclei could dedifferentiate and fully support embryonic development. This theory needed to be tested in mammals. The reports of Illmensee and Hoppe and Modlinski had emphasized the high death rate for zygotes whose nuclei were removed and replaced with a nucleus from another cell. Therefore, throughout the 1980s, research teams sought new techniques. Because mammalian eggs are so much smaller than amphibian eggs, and blastomere nuclei are so much larger than somatic cell nuclei, the techniques of nuclear transplantation into mammalian eggs had to be perfected.

Cell fusion. To avoid puncturing the plasma membrane of the mammalian egg, Davor Solter and his colleagues at the Wistar Institute in Pennsylvania took advantage of the property of some proteins, most notably those expressed on the surface of Sendai virus, to cause two plasma membranes to fuse together. In 1983, they reported a series of experiments in which they removed both pronuclei from zygotes and replaced them with pronuclei from another zygote. The two pronuclei were removed at the same time with a glass pipette similar to the one described by Illmensee, and new pronuclei were introduced into the enucleated zygote by a membrane fusion technique (Figure 10.2). The pronuclei to be transplanted into the egg were drawn up into the glass micropipette, exposed to a small amount of inactivated Sendai virus, then placed into the **perivitelline space** between the zona pellucida and the zygote plasma membrane (Figure 10.2). The manipulated zygote was returned to the incubator and fusion took place within approximately 1 hour. The developmental potential of the reconstructed zygotes was almost the same as nonmanipulated controls.

Perivitelline space: the area between the egg plasma membrane and the zona pellucida.

Using this technique, McGrath and Solter demonstrated that nuclei from cleaving blastomeres could also be fused to enucleated zygotes by inactivated Sendai virus (Figure 10.2). In 1984, they published a report that refuted the work of Illmensee and Hoppe in 1981. Their studies demonstrated that 99% of zygotes could

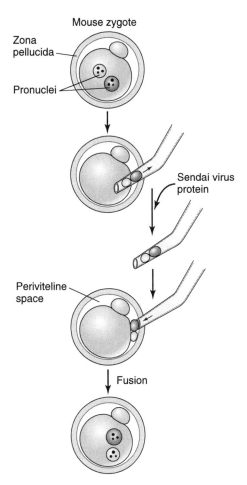

Figure 10.2   Transplantation of mouse pronuclei by membrane fusion.

survive removal of both pronuclei. As before, if fused with pronuclei from another zygote, 95% of the reconstructed zygotes developed to blastocysts in culture. If fused with the nucleus from a Two-Cell, 81% cleaved once or twice, but only 15% of those developed to blastocysts. If fused with the nucleus from a Four-Cell, 95% cleaved once or twice, but none developed to a blastocyst. If fused with the nucleus from a morula, all cleaved once or twice, but none developed to blastocysts. Finally, if fused with an ICM cell, an experiment analogous to that of Illmensee and Modlinski, all reconstructed zygotes cleaved once or twice, but none developed to blastocysts.

These results were so different from those reported in 1981 by the other investigators that McGrath and Solter repeated a series of experiments in which nuclei were injected directly into the zygote cytoplasm instead of using their fusion technique.

The outcome did not change. None of the enucleated zygotes injected with ICM nuclei (72 out of 231 zygotes manipulated survived the reconstruction procedure) developed past one or two cleavages. They speculated that Illmensee had not removed all remnants of the pronuclei from the zygotes he enucleated, and that those remnants supported the subsequent development to blastocysts and offspring that he observed.

McGrath and Solter concluded their 1984 report with the sentence "Differential activity of maternal and paternal genomes, and the results presented here, suggest that the cloning of mammals by simple nuclear transfer is biologically impossible." This statement influenced scientific thinking for several years.

Subsequent work with cow zygotes by Jim Robl in the University of Wisconsin laboratory of Neal First supported the McGrath and Solter 1984 report. Robl substituted **electrofusion** for the Sendai virus-induced fusion practiced by McGrath. The principle behind electrofusion is similar to sendai virus-mediated fusion, but the plasma membranes are caused to fuse together by a small electrical jolt (Figure 10.3). Robl fused pronuclei with enucleated cow zygotes to derive reconstructed embryos. Two reconstructed embryos were transferred to pregnant recipients and two calves were born, thus demonstrating that the process of manipulation and electrofusion did not significantly alter developmental potential. However, nuclei from two-, four-, or eight-cell embryos fused to enucleated zygotes did not produce embryos that could develop further. This work was published in 1987 and supported the work with mouse zygotes reported by McGrath and Solter that revealed that the developmental potential of mammalian blastomere nuclei is restricted relatively early.

However, the inability of blastomere nuclei to support embryonic development was not consistent with the clear pluripotent nature of ICM cells, and investigators continued to probe early embryonic systems for better answers. One such study was also published in 1987 by Yukio Tsunoda working in the laboratory of Professor Sugie. Tsunoda and his colleagues transferred nuclei from Four-Cells and Eight-Cells into enucleated Two-Cells and obtained not only improved development in laboratory culture, but the birth of live offspring (Figure 10.4). This suggested that cytoplasmic factors in zygotes were incompatible with transferred nuclei, and supported the concept that the nuclei from eight-cell stage blastomeres could still direct full development of new offspring. The problem seemed to lie in some incompatibility between the transplanted nuclei and the cytoplasm of the zygote.

By the late 1980s, the power of the cell cycle regulators described in Chapter 2 over nuclear events had become widely appreciated. Developmental biologists began to explore the relationship between the stages of the cell cycle of the zygote and the nucleus being transplanted.

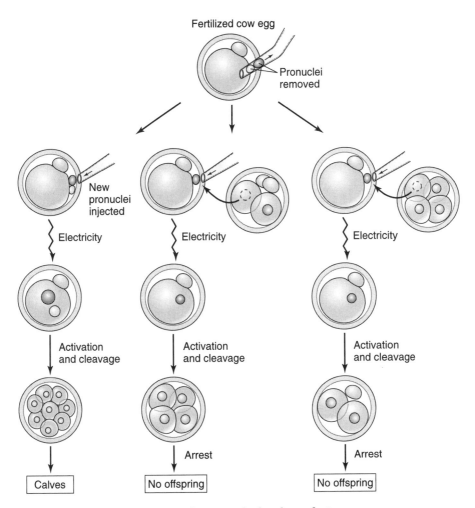

Figure 10.3 Transplantation of cow nuclei by electrofusion.

Nuclear Transplantation and the Cell Cycle of Fertilized Eggs

One of the earliest studies that focused on the role of the cell cycle in successful blastomere cleavage following nuclear transplantation was conducted by Smith, Wilmut, and Hunter at the Edinburgh Research Station in Scotland. Mouse embryos were their experimental system. They reported in 1988 that cross transfer of pronuclei from one zygote to another resulted in a higher rate of blastocyst development (94%) if the donor and recipient zygotes were synchronized with respect to fertilization. In the same study, they observed that the timing of the transplantation of Two-Cell nuclei to enucleated zygotes was more complicated. Both the **karyoplast** (the Two-Cell nucleus) and

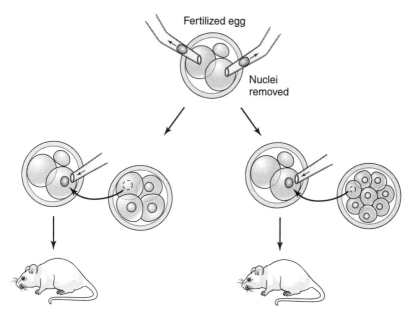

Fertilized egg

Nuclei removed

Figure 10.4 Transplantation of nuclei from Four-Cell and Eight-Cell into enucleated Two-Cell.

Cytoplast: the egg cell remaining after the removal of all chromosomes.

the **cytoplast** (the enucleated zygote) influenced cell cleavage and successful development to blastocysts. Overall, reconstructed embryos from Two-Cell nuclei at later stages of the Two-Cell cycle (during the long G2 phase described in Chapter 6) transplanted into late stage (G2) zygotes had the greatest potential to develop to blastocysts. This observation suggests that nuclei that had completed S phase and passed through the checkpoint for DNA replication were better prepared to resume division in a strange cytoplasm, especially if the cytoplasm were also at the G2 stage of the cell cycle.

A follow-up study published in 1990 by Smith, Wilmut, and West examined the outcome of fusing early and late karyoplasts from zygotes and Two-Cells to early and late zygote cytoplasts. Early pronuclei and Two-Cell nuclei (probably in early S phase, since we now know that the G1 phase of both pronuclei and Two-Cells is short) fused with late zygote or Two-Cell cytoplasts delayed cleavage for a few hours. This agrees with the notion that nuclei in S phase can complete DNA synthesis in the presence of G2 cytoplasm because the nucleus has gathered within itself all the necessary enzymes and supporting elements for DNA replication. Moreover, G2 cytoplasm has little influence over the nucleus until S phase is completed. In contrast, late pronuclei and Two-Cell nuclei (in G2) fused with early enucleated zygote and Two-Cell cytoplasts (probably in the G1/S transition or S phase) resulted in embryos that did not cleave earlier than expected, according to the age of the cytoplasm. This important observation is consistent with the known cyclin/Cdc control of the G2 to M phase transition discussed in Chapter 2. Many late Two-Cell karyoplasts transferred into early zygote cytoplasts failed to undergo cleavage at all, suggesting that the long G2 phase of the mouse Two-Cell is required for a specific interaction between the nucleus and the cytoplasm in this species. Whether or not this is true for other species is not known with certainty.

Transplantation of Embryonic Nuclei into Unfertilized Eggs

Of particular importance to the development of human therapeutic cloning procedures has been the work done in animals developing the methods to use unfertilized eggs to receive transplanted nuclei. Unfertilized eggs are not only far easier to acquire as an experimental model, but they avoid the ethical concerns associated with human eggs fertilized by sperm. This approach necessitated an improved understanding of egg physiology, including egg activation without sperm. Termed parthenogenesis (discussed in Chapter 4), egg activation requires both overcoming the cellular controls responsible for maintaining arrest at metaphase II (see sidebar, Chapter 4), and then deploying the egg stores responsible for several rounds of chromosome duplication and blastomere cleavage. In addition, since the goal of nuclear transplantation experiments was to test the potential of differentiated nuclei to support embryonic development, the transplanted nucleus could need to undergo remodeling to a dedifferentiated state before cleavage was initiated. A tall order.

Although some reproductive biologists pursued studies of nuclear transplantation into eggs to more fully understand cell differentiation, others were interested in developing methods for cloning animals, especially valuable farm stock. In this regard, multiple uses of the term clone has created some confusion. As noted by early developmental biologists (see sidebar, Chapter 7), separation of early cleaving blastomeres could result in the development of multiple offspring. Twins created by separation of Two-Cells, for example, are relatively common in many species. Steen Willadsen reported a number of studies throughout the 1980s aimed at producing identical twins and triplets from cow and sheep embryos that had been manually bisected or quadrasected (Figure 10.5). Some experiments were successful, and twins and quadruplets developed from single embryos. Such offspring are, strictly speaking, genetic clones of each other.

The other use of the term clone refers to individuals who result from the transfer of a nucleus from an existing individual (including adults) into an egg that is then induced to initiate embryonic development. It is this procedure that McGrath and

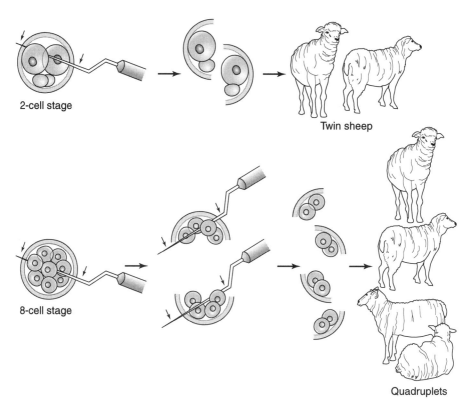

2-cell stage

Twin sheep

8-cell stage

Quadruplets

Figure 10.5 Multiple offspring ("clones") produced from single embryos.

Solter deemed "…not biologically possible." As the following sections will describe, however, their prediction was proven wrong.

Willadsen also conducted early studies of blastomere fusion with unfertilized eggs using sheep eggs and embryos. He reported his findings in 1986. His basic procedure was to bisect unfertilized eggs to remove their genetic information, then fuse the half of the egg with no chromosomes to whole blastomeres from sheep morulae at the 8- to 16-cell stage. Although the success rates were low, a few lambs were born from the reconstructed half-eggs. This work demonstrated not only the feasibility of using unfertilized eggs to give rise to viable offspring, but the power of egg cytoplasm to remodel nuclei.

Similar experiments were being conducted at the same time by Roger Prather in the Wisconsin Laboratory of Neal First. Unfertilized oocytes were tested as suitable recipients for blastomere nuclei. A work published in 1987 described how two calves had been born from embryos reconstructed by electrofusion of morula stage blastomere nuclei with enucleated, unfertilized eggs. Step one (Figure 10.6) removed

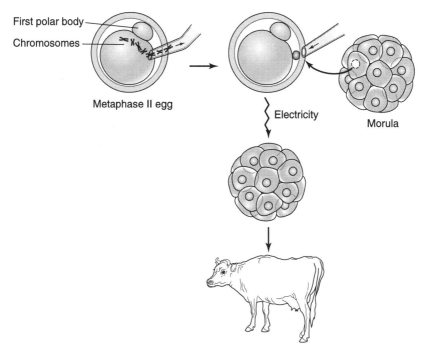

Figure 10.6 Transplantation of cow morula nuclei into unfertilized egg.

the egg's chromosomes. Although this step was termed *enucleation* by the authors, strictly speaking it was chromosome removal since metaphase II eggs do not have a nucleus. Following removal of the egg chromosomes, the resulting egg cytoplast was fused with nuclei from morula stage cow embryos using electrical pulses, as previously reported by Robl. The electrical pulses not only fused the blastomere nucleus with the egg cytoplast, they also served to activate the egg so it would initiate cleavage into blastomeres.

Although the success rates were low (fewer than 1% of fused eggs resulted in calves), these results raised the provocative possibility of endlessly cloning one cow (Figure 10.7). Unfertilized cow eggs are readily available from ovaries of slaughterhouse animals, thus markedly easing the logistics of needing fertilized eggs. Theoretically, each of the 16 blastomeres of a cow morula could be fused with enucleated, unfertilized cow eggs, giving rise to 16 identical morulae. Some of those could be transferred to the uteri of pregnant cows for gestation, others could be preserved by freezing them in liquid nitrogen, and others could serve as the source of blastomeres for the next round of cloning. This possibility was met with enthusiasm by people wishing to produce genetically identical offspring for a variety of reasons, including livestock production, rescue of endangered

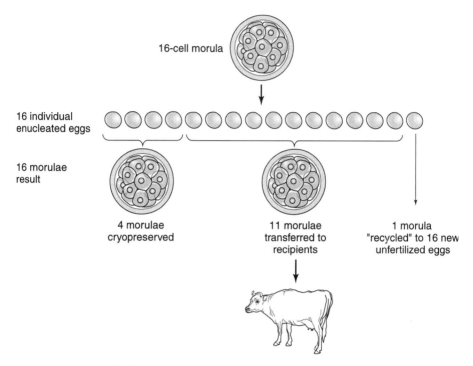

Figure 10.7 Endlessly cloning one cow by recycling morula nuclei.

species, and drug testing and other types of research. Production companies sprang up all over the world. However, the work was greeted with horror by people concerned about eugenics and the creation of an overabundance of genetically identical individuals, thus biasing the gene pool and by-passing essential processes of natural selection.

Importantly, the theoretical possibility was not realized. For reasons still not fully understood, second and third generations of cloned blastomeres had a markedly reduced potential to develop to offspring. The success rate was well below 1%, making the process economically unfeasible. Unraveling the biology behind this failure will provide important insights into the subtle, but highly significant, processes that underlie successful embryologic development. The work will take years and major financial resources. As of this writing, research in this area is proceeding slowly.

Nonetheless, the possibility of obtaining multiple, genetically identical individuals from morulae has remained attractive for research purposes, even if only one generation were possible. This work continues for research purposes in laboratories throughout the world.

Transplantation of Somatic Cell Nuclei into Unfertilized Eggs

The ability to generate cleaving blastomeres from nuclei transplanted into unfertilized eggs of several species greatly improved the opportunity to continue to explore the developmental potential of nuclei from a variety of cell types. The goals of the continuing research included not only basic developmental biology questions related to gene expression and nuclear reprogramming, but also an interest in applying the emerging genetic manipulation techniques demonstrated in mice to larger animals. For example, as discussed in Chapter 8, mouse ES cells become incorporated into the ICM when injected into the blastocoel cavity and subsequently contribute to the tissues of the developing offspring. Since mouse ES cells can be genetically engineered in a variety of ways, including homologous recombination with altered gene sequences (Chapter 8), it is theoretically possible to derive an endless variety of genetically altered mice. Such mice provide powerful probes into the function of specific genes and interaction with other genes. Perhaps even more provocative is the opportunity to correct mutated genes in the genetic background of the animal, or add new genes. On the one hand, this line of investigation again raises the specter of eugenics and is to be pursued with due caution; on the other hand, the amount of suffering that could be alleviated by correcting known genetic defects cannot be ignored.

In addition, there is the possibility of genetically engineering animals to express proteins of therapeutic value. One example of this is the expression of tissue plasminogen activator (TPA) in the milk of goats. TPA has been shown to improve the outcome of people suffering heart attacks if administered quickly. The gene for human TPA has been added to the germ line of some goats by injecting DNA constructs containing the gene into the pronuclei of fertilized goat eggs (as described in the sidebar, Chapter 5). Expression of TPA in goat milk apparently has no side effects to the animal. The process of deriving such an animal through transgenic technology is expensive and lengthy. Once such an animal is obtained, there would be great advantage in being able to clone her from her somatic cells. Only in this way would exact genetic replicas of her be possible.

By the late 1980s, groups studying nuclear transplantation began to explore the ability of ICM cells fused with unfertilized eggs to direct blastomere cleavage. In 1989, Smith and Wilmut reported the birth of lambs developed from embryos reconstituted from either 16-cell blastomeres or ICM cells fused with unfertilized sheep eggs (Figure 10.8). This was one step closer to cloning an animal from a somatic cell. What was needed was a line of sheep ES cells derived from a sheep blastocyst.

Two research teams (G. B. Anderson, and Modlinski and Illmensee) attempted to clone mice from embryos reconstructed by transplantation of mouse ES cell nuclei into unfertilized mouse eggs in the early 1990s, but they failed to produce live offspring. This work reinforced the concept that only early cleaving blastomere nuclei were dedifferentiated enough to fully support embryonic development and that, although cultured

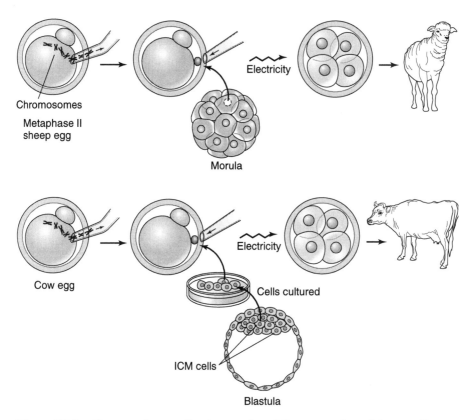

Chromosomes

Metaphase II
sheep egg

Morula

Electricity

Cow egg

Electricity

Cells cultured

ICM cells

Blastula

Figure 10.8 Sheep and cow offspring produced from morula and inner cell mass (ICM) cell nuclei.

ES cells could contribute to most tissues when injected into blastocysts, they were not totipotent. It appeared as though there was a reduction in totipotency when ICM cells were adapted to long-term culture as ES cells.

In 1993, Sims and First reported the first calves born from embryos reconstituted from cultured ICM cells fused to unfertilized cow eggs. The ICM cells were cultured for 1 to 4 weeks at the time of nuclear transfer (Figure 10.8). An established line of ES cells had not been derived from cow blastocysts, a situation that still exists today. Nonetheless, this landmark experiment proved that it may be possible to obtain offspring from ICM cells cultured long enough to derive at least a few cells with new gene inserts through homologous recombination.

This work was followed in sheep by Keith Campbell and associates, working in the laboratory of Ian Wilmut. This team derived cell lines not from blastocysts, but from fetal sheep tissues. As had been demonstrated with the cow ICM cells, this team reported in 1996 that it was possible to obtain live offspring from embryos constructed by transplanting nuclei from cultured fetal cells.

Then came Dolly. In 1997, Ian Wilmut and associates reported the birth of a ewe from an embryo constructed from the nucleus of an adult sheep breast cell line that had been in culture for several generations. The prediction that "…cloning by simple nuclear transfer is biologically impossible" had been resoundingly disproven.

In summary, 20 years ago, scientists at the Jackson Laboratories reported the birth of mice from embryos constructed of nuclei from blastocyst ICM (but not trophoblast) cells transplanted into zygotes whose pronuclei were removed. Because the work could not be repeated, it was called into question. Subsequent studies with zygotes and early cleaving blastomeres suggested that the potential for mammalian cells to direct embryonic development was restricted within the first few cleavage divisions, in contrast to amphibian cells. By the late 1980s, scientists had begun to explore the feasibility of constructing embryos from unfertilized eggs, rather than from zygotes. Successes with this approach accelerated research, particularly with large animals, because of the ready availability of cow and sheep eggs from slaughterhouses. The theoretical possibility of endlessly cloning one cow embryo met with such reduced success at each round of cloning, for reasons still not explained, that the approach was abandoned. Each year throughout the early 1990s brought improved success with reconstructing embryos from unfertilized eggs and cells at increasingly advanced stages of differentiation, until finally Dolly was born, reconstructed with the nucleus of a long-term tissue culture cell line. This opened the door to cloning all species from adult somatic cells and launched the ethical debates about eugenics that rage today.

Animal Cloning

In 1998, a research team at the University of Hawaii cloned mice from granulosa cell nuclei and Jim Robl's team at the University of Massachusetts cloned cows from fetal fibroblasts that had been rendered transgenic by homologous recombination with a DNA construct containing a gene that encoded a protein easily distinguishable from cow proteins. These breakthroughs confirmed the feasibility of carrying out specific genetic manipulations in cattle, and opened the possibility of engineering livestock to produce therapeutics in large quantities, such as in milk. In 1999, live goats were produced from embryos constructed from fetal fibroblasts, and in 2000, pigs were produced by procedures similar to the goat. Thus, as of this writing, offspring have been cloned from somatic cell nuclei from cows, sheep, goats, pigs, mice, rabbits, and cats.

Importantly, however, the percentage of live offspring from reconstructed embryos remains low, on the order of 1 or 2%. In addition, some die within the first few weeks of life. The reasons for this are not clear, but do not appear to be the result of chromosomal anomalies because the offspring studied, including those that died, seem to have a normal karyotype. This suggests that the deaths resulted from failures in epigenetic controls, such as those involved in imprinting.

Nonetheless, the cloning successes have demonstrated the ability of eggs to dedifferentiate somatic cells into totipotent embryonic cells. This has established a fundamentally new biologic principle. Whether or not animal cloning can be made more efficient and healthy for the offspring, the power of the discovery to derive pluripotent stem cells for therapeutic purposes has been firmly established.

Challenges in Human Therapeutic Cloning

As discussed in the preceding chapters, nuclear reprogramming and egg activation must both be accomplished to bring about the self-sustained cleavages of early blastomeres derived from transplantation of somatic cell nuclei into enucleated human eggs (see sidebar). Until methods are developed to derive pluripotent stem cell lines directly from blastomeres, stem cells will need to be derived from ICM cells, in a manner similar to that of the ES cells of the mouse. Thus, for the foreseeable future, human therapeutic cloning will depend on the ability to activate enucleated eggs with transplanted nuclei in a manner analogous to animal cloning.

The cell cycle stage of the somatic cell nucleus being transferred has been found to play a role in successful subsequent cleavages. Since the activated egg is completing M phase and immediately entering G1, the somatic cell nuclei will be most synchronous if they are arrested in G1 of their cell cycle. Several methods are available to bring about a G1 arrest of tissue culture cells, including withdrawing the growth factors present in the serum that drives cell division. If this is accomplished, the G1 phase of the activated egg may be too short to allow reprogramming of the somatic cell nucleus, as well as recruitment into S phase, as directed by the cell cycle proteins. This is an area of active investigation.

Taken together, work to date has revealed that **nuclear reprogramming** involves decreased expression of cell-type specific genes, for example, hemoglobin, and increased ease of expression of a broader range of genes. Importantly, some gene products function to inhibit the expression of other genes. It is possible that a fundamental feature of reprogramming a somatic cell nucleus to support embryonic development is broad suppression of all genes leading to cell differentiation coupled with broad expression of all genes involved in DNA replication and cell division. Given that this can be accomplished, the next requirement would be to invoke the systematic expression of groups of genes in the correct order to bring about cell differentiation.

Nuclear reprogramming: the process of changing gene expression from a differentiated cell state to an undifferentiated cell state to restore pluripotency.

SOMATIC CELL NUCLEAR TRANSFER IN HUMANS

Pronuclear and Early Embryonic Development

Jose Cibelli completed veterinary school in Argentina. Rather than immediately entering veterinary practice, he decided to pursue his interest in cloning cattle by enrolling in graduate school in the Animal Science program at the University of Massachusetts. He married his high school sweetheart and convinced her to travel to Massachusetts with him. It was inevitable that his interests would lead him to the laboratory of Jim Robl.

Jose honed his microscopy and micromanipulation skills by conducting experiments with both rabbit and cow eggs. In 1998, he published the studies that announced the birth of the first transgenic calves. The low efficiency of production of offspring, and the relatively high frequency of perinatal death, were discouraging, but Jose had already realized that the real benefit from nuclear transfer embryos was the potential to derive stem cells for therapeutic purposes.

During his graduate education, a small biotechnology company was formed. The patents that arose from the nuclear transfer experiments had attracted investors in the technology. Advanced Cell Technology was formed to facilitate continuing the work with investor financing. The goal was to develop the methods for deriving human stem cells from somatic cell nuclei (such as skin fibroblasts) transplanted into enucleated eggs. The logistical problems inherent in acquiring human eggs for research purposes were staggering, and many of the methods of enucleation, nuclear transplantation, and egg activation had not been tested in humans.

Advanced Cell Technology was acquired by Michael West, a scientist with experience in running biotechnology companies. West had previously founded Geron corporation, a California biotechnology company focused on regenerative medicine. Dr. West had conducted seminal experiments on the enzyme telomerase, which maintains the ends of chromosomes. Shortening of the stretch of nucleic acids at the ends of chromosomes is thought to play a role in cellular aging. Dr. West's passion to develop human therapeutic stem cells matched that of Jose Cibelli.

The for-profit nature of the company brought sharp criticism from critics of the proposed work, to conduct nuclear transplantation into human eggs, which would then be activated to initiate self-sustained cleavages to the blastocyst stage. The initial steps were akin to cloning a human. The company and its scientists buffeted the public outcry and compiled an ethics advisory board to provide outside guidance for recruiting women to donate eggs for research. Sharp criticism was then leveled at the ethics board as well. Congress threatened to outlaw all forms of human nuclear transplantation into eggs for the purposes of deriving blastocysts for research.

Nonetheless, carefully following the lengthy guidelines of the ethics board, Jose Cibelli and his colleagues were able to stimulate a few human eggs into initiating cleavage divisions, both without and with nuclear transplantation. The work was published in 2001. A total of 71 eggs were recovered from the ovaries of seven women research volunteers. Twenty-two were activated parthenogenically to test the protocols for artificial egg activation. Ninety percent cleaved, 30% of which developed to at least the early blastocyst stage. This result predicted that it would be possible to derive human ES cells from parthenogenically activated blastocysts. If such ES cells were proven to be histocompatible with the woman from which they were derived, this, in itself, offered an opportunity to develop stem cell therapy for ovulating women.

In addition, 11 of the eggs underwent nuclear transplantation with cultured fibro-blast nuclei before they were activated. Although some nuclear remodeling was evident, none of the eggs cleaved. Eight of the eggs underwent nuclear transplantation with granulosa cell nuclei from the egg donors. Three of those eight cleaved beyond two cells.

Although the numbers were small, the road to human therapeutic cloning had begun.

Additional Readings

Baquisi, A., Behboodi, E., Melican, D. T., Pollock, J. S., Destrempes, M. M., Cammuso, C., Williams, J. L., Nims, S. D., Porter, C. A., Midura, P., Palacios, M. J., Ayres, S. L., Denniston, R. S., Hayes, M. L., Ziomek, C. A., Meade, H. M., Godke, R. A., Gavin, W. G., Overstrom, E. W., and Echelard, Y. (1999). Production of Goats by Somatic Cell Nuclear Transfer. *Nature Biotechnol* 17: 456–461.

Campbell, K. H. S., McWhir, J., Ritchie, W. A., and Wilmut, I. (1996). Sheep Cloned by Nuclear Transfer from a Cultured Cell Line. *Nature* 380: 64–66.

Illmensee, K., and Hoppe, P. C. (1981). Nuclear Transplantation in *Mus musculus*: Developmental Potential of Nuclei from Preimplantation Embryos. *Cell* 23: 9–18.

Kuroiwa, Y., Kasinathan, P., Choi, Y. J., Naeem, R., Tomizuka, K., Sullivan, E. J., Knott, J. G., Duteau, A., Goldsby, R. A., Osborne, B. A., Ishida, I., and Robl, J. M. (2002). Cloned Transchromosomic Calves Producing Human Immunoglobulin. *Nature Biotechnol* 20: 889–894.

McGrath, J., and Solter, D. (1984). Inability of Mouse Blastomere Nuclei Transferred to Enucleated Zygotes to Support Development in Vitro. *Science* 226: 1317–1319.

Modlinski, J. A. (1981). The Fate if ICM and Trophectoderm nuclei transplanted to fertilized mouse eggs. *Nature* 292: 342–343.

Sims, M., and First, N. L. (1993). Production of Calves by Transfer of Nuclei from Cultured Inner Cell Mass Cells. *Proceedings of the National Academy of Sciences USA* 90: 6143–6147.

Willadsen, S. M. (1986). Nuclear Transplantation in Sheep Embryos. *Nature* 320: 63–65.

Yanagimachi, R. Cloning: Experience from the Mouse and Other Animals. (2002). *Mol Cellular Endocrinol* 187: 241–248.

CHAPTER 11

Stem Cell Differentiation

I have not failed. I've just found 10,000 ways that won't work.
Thomas Alva Edison

OVERVIEW

As discussed in previous chapters, stem cells are capable of unlimited cycles of growth and division into genetically faithful daughter cells, and can be induced to differentiate into a wide variety of cell types. It has been known for some time that there are multiple classes of stem cells, some of which are already committed to a tissue type (for example, skin stem cells). Because the field is so new, the nomenclature is loose and stem cells are generally classified as either "adult," "fetal," or "embryonic," as described in Chapter 1. This classification refers simply to their source and does not accurately reflect their level of commitment or their differentiation potential. In many instances, commitment and differentiation potential are not known with certainty because the work is ongoing. As a general rule, adult stem cells are thought to be more committed than fetal stem cells, which are in turn more committed than embryonic stem cells. As the stem cell field matures, more accurate nomenclature will undoubtedly develop.

The overarching consideration for stem cells is how to get them from the laboratory to clinical treatments. Much needs to be done. This chapter will outline what is known and what is not known to accomplish this.

Adult Stem Cells

One suggestion for a more accurate term for adult stem cells has been *multipotent adult progenitor cells* or MAPCs. This term includes the emerging notion that some adult stem cells may actually be able to contribute to tissues other than their tissue of origin. Based on embryology (Chapter 9), one line of thinking about adult stem cells has been that they would stay faithful to their tissue of origin, or at the least, to their tissue of embryonic origin. For example, stem cells from adult bone marrow obviously give rise to all the blood cell lineages (see sidebar, Chapter 1). In addition, they could perhaps give rise to other cell types from tissues of mesodermal origin, but not from ectodermal or endodermal origin. At least for mice and rats, this long-held view has now been disproven in the laboratory of Catherine Verfaillie at the University of Minnesota. In a series of elegant experiments, her research team isolated mouse MAPCs (mMAPCs) from mouse bone marrow and first demonstrated that a single cell was capable of continually dividing into a large number of identical daughter cells (termed **clonal expansion**). In addition, when those cells were exposed to appropriate growth factors in laboratory culture systems, they developed the characteristics of both neuronal cells (ectodermal) and liver cells (endodermal).

> **Clonal expansion:** one cell giving rise to millions of identical cells.

Moreover, single mMAP cells were injected into mouse blastocysts (see sidebar, Chapter 8) to generate chimeric offspring. If the single mMAP cell could divide in the ICM along with the blastocyst ICM, then in theory it could contribute to a variety of tissues in offspring along with the ICM cells themselves. The single mMAP cell injected into the blastocysts contained a **reporter gene** to allow it to be identified. The injected blastocysts were returned to the uteri of recipients for gestation and birth. The tissues of such offspring were examined for evidence of the single injected mMAP cell. Although only a small percentage of offspring were successful chimeras, those that were contained mMAP cells in brain, retina, liver, intestine, kidney, spleen, bone marrow, and skin (Figure 11.1). These findings suggest either that the bone marrow mMAPs were intrinsically capable of contributing to tissues of mesodermal, endodermal, and ectodermal origin, or that they acquired that ability during their time in laboratory culture.

As an additional test of the potential for the mMAPs to contribute to adult tissues, they were injected into blood veins of adult mice to determine if they could establish themselves in any tissues. After a few months, tissues of the injected mice were examined for evidence of the mMAPs; they were found in blood, bone marrow, and spleen, as well as lung, liver, and intestine (Figure 11.2). These tissues are known to have their own population of stem cells. Importantly, the mMAPs were not found in skeletal muscle, heart muscle, brain, or kidney, all tissues known to be deficient in their own populations of stem cells. These findings suggest that there are

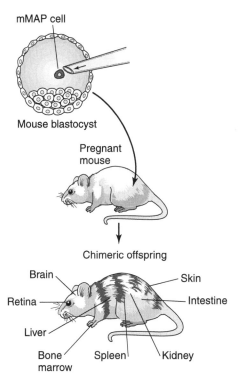

Figure 11.1 Tissue distribution of single mouse MAP (mMAP) cell injected into a mouse blastocyst.

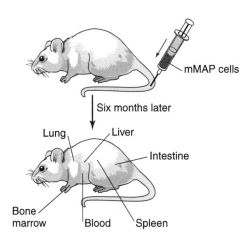

Figure 11.2 Tissue distribution of mouse MAP (mMAP) cells injected into the tail vein of an adult mouse.

rigorous control mechanisms for controlling stem cell populations in organs that may need to be overcome for stem cell therapy to be clinically effective.

These landmark findings about adult bone marrow stem cells, from the laboratory of Catherine Verfaillie, need to be confirmed in other laboratories. Alternative explanations, such as fusion with embryonic cells, have already been put forth by other experts in the field. This is an active area of investigation, with new findings reported weekly. Thus, time will tell if bone marrow cells from adult animals contain pluripotent stem cells capable of contributing to a wide variety of tissues across all embryonic cell layers of origin.

Fetal Stem Cells

Three sources of fetal stem cells have been investigated: trophoblast stem cells (TS cells), primordial germ cells, and fetal tissue stem cells. As with some other types of stem cells, the principal difficulty in obtaining long-term, self-renewing populations of TS cells was the inability to prevent them from differentiating. This is an important observation, and seems to also apply to many, if not all, stem cells. Under many laboratory culture conditions, stem cells spontaneously differentiate and stop dividing. This suggests that differentiation is the default mode for cells, which, in many respects, provides a natural limitation to the size of organs and a barrier to tumor formation. Therefore, the environment that supports an uncommitted state must promote expression of those genes whose products inhibit differentiation. In tissues, the environment must also inhibit cell division unless new cells are needed, such as described for the skin injury in Chapter 2.

Therefore, a fundamentally important development that led to successful laboratory culture of TS cells was the discovery, in the laboratory of Janet Rossant in Toronto, that the growth factor **fibroblast growth factor 4** (FGF4, which uses heparin as a co-factor) was as necessary as the fibroblast feeder layer to maintain trophoblast cells in an undifferentiated state. This has allowed the development of continually renewing cell lines of TS cells for study. Injection of TS cells into blastocysts has revealed that they contribute exclusively to trophoblastic tissues in developing mouse fetuses, thus limiting their usefulness as therapeutic cells for other tissues. Additional studies of TS cells has revealed, however, that conditioned medium from the fibroblast feeder layers will substitute for the fibroblast cells themselves. This observation opens the way to identify those factors contributed by the feeder layer that are necessary to maintain the undifferentiated state of the TS cells. Such factors may also play an important role in maintaining the undifferentiated state of other stem cells.

FGF4: *fibroblast growth factor number 4.*

Circumventing the need for a feeder layer to promote division and inhibit differentiation of stem cells is important to stem cell therapy. All lines of human ES cells currently available were derived atop a feeder layer of mouse fibroblasts. Thus, the ES cells were continually exposed to mouse cells, and there is concern that those mouse cells may spontaneously express unusual pathogens, such as endogenous retroviruses. Although this concern is theoretical at this time, there is a need to prove the safety of mouse fibroblast feeder layers for stem cells being cultured for therapeutic purposes.

Stem cells from human primordial germ cells, termed human embryonic germ (EG) cells, were first isolated in the laboratory of John Gearhart in 1999. Their origin was the gonadal ridge (see Chapter 9) of 5- to 9-week fetuses from elective abortions. Although highly controversial, the characteristics of EG cells may provide valuable clues about maintaining the fully undifferentiated state of human embryonic stem cells from less controversial sources. A comparison of their gene expression and culture characteristics to other sources of human embryonic stem cells is listed in Table 11.1. It may be an important feature of their therapeutic developmental potential that they do not grow into teratomas in immune compromised mice.

Human Embryonic Stem Cells

Human ES (hES) cells are without a doubt the most pluripotent of all stem cells and, for that reason, hold the most therapeutic promise. It is important to distinguish the several types of hES cells, their therapeutic potential, and their status as of this writing.

As described in Chapter 8, mouse ES cells paved the way for isolation of hES cells (see sidebar), and some important similarities and differences between mouse ES cells and human ES cells have emerged, as listed in Table 11.2. The importance of each of these characteristics to both of the underlying properties of stem cells, immortality and pluripotential for differentiation, is an active area of research in several laboratories throughout the world. For a variety of reasons, none of the hES cell lines currently available may prove useful for therapeutic purposes, but studies of their characteristics will be invaluable guides to hES cells destined for therapeutic purposes. For purposes of discussion in this text, the ES cell lines in existence at this writing will be designated **hFES cells** (human fertilized embryonic stem cells).

The work with hFES cells is just beginning, and results to date are very promising. First, although there are major differences among the various hFES cells available with respect to culture characteristics and rate of self renewal, they all appear capable of proliferation for years and hundreds of population doublings without becoming aneuploid. It is an extremely important characteristic that their chromosomal composition remains stable throughout many cell cycles.

In addition, work in laboratories in the U.S., Israel, and Australia have all demonstrated that hFES cells are capable of differentiating into cell types derived from all

Table 11.1

Comparison of Human Pluripotent Stem Cells

Characteristic	Embryonic Stem Cells from Eggs Fertilized by Sperm	Embryonic Germ Cells from Human Fetal Gonadal Ridge	Embryonic Carcinoma Cells from Human Tumors
Express stage-specific embryonic antigens (SSEA)-1, -3, -4	SSEA-3, -4	SSEA-1, -3, -4	SSEA-3, -4
Express tumor rejection antigen (TRA)-1	Yes	Yes	Yes
Express Oct-4	Yes	Yes	Yes
Express alkaline phosphatase	Yes	Yes	Yes
Express telomerase	Yes	Unknown	Yes
Factors that aid in self renewal	Feeder cell layer plus either serum or bFGF	LIF, bFGF, and forskolin	Not well characterized
Growth characteristics in culture	Form flat, loose aggregates	Form rounded, multilayer clumps	Form flat, loose aggregates
Can form embryoid bodies	Yes	Yes	Yes
Grow into teratomas in nude mice	Yes	No	Yes

three embryonic cell layers: ectoderm, mesoderm, and endoderm. Studies demonstrating that this is also true in an animal setting, as well as a laboratory culture setting, involve injecting hFES cells into mice with no immune system. Under these circumstances, the hFES cells give rise to teratomas. Analyses of the cell types contained

Table 11.2

Comparison of Human and Mouse Embryonic Stem Cells

Characteristic	Human Embryonic Stem Cells from Eggs Fertilized by Sperm	Mouse Embryonic Stem Cells from Eggs Fertilized by Sperm
Express stage-specific embryonic antigens (SSEA)-1, -3, -4	SSEA-3, -4	SSEA-1
Express tumor rejection antigen (TRA)-1	Yes	No
Express Oct-4	Yes	Yes
Express alkaline phosphatase	Yes	Yes
Express telomerase	Yes	Yes
Factors that aid in self renewal	Feeder cell layer plus either serum or bFGF	LIF and other factors acting through gp 130 receptor
Growth characteristics in culture	Form flat, loose aggregates	Form tight, rounded, multilayer clumps
Can form embryoid bodies	Yes	Yes
Grow into teratomas in nude mice	Yes	Yes
Can contribute to chimeras	Unknown	Yes

within the teratomas have revealed structures resembling gut epithelium (endodermal in origin), smooth and striated muscle cells (mesodermal in origin), and neural and squamous epithelium (ectodermal in origin).

For obvious ethical reasons, some of the experiments conducted with single mMAPs cannot be repeated with human cells. A single hFES cell cannot be microinjected into a human blastocyst before transfer to the uterus to discover which tissues of the resulting baby have become chimeric for the injected cell. Since this has already been demonstrated in animal models, however, it will be more important to determine if the specific disease being addressed with stem cell therapy is ameliorated following treatment with stem cells cultured and differentiated for that specific purpose. Thus, the fact that a given stem cell line can contribute to a chimera is as fundamentally important to prove for hFES cells as for animal ES cells.

As will be described in Chapters 12 and 13, hES cells used for therapies will need to be free of pathogens, tissue compatible with the recipient, capable of continual renewal in culture, and well controlled in the tissue of the donor so as to not spontaneously form tumors. These considerations support the emerging view that hFES cells derived from eggs fertilized by sperm or from fetal tissues, including fetal germ cells, are not the ideal hES cells for therapeutic purposes.

Human Parthenote Stem Cells

Two other types of hES cells may hold more promise: those obtained from parthenogenically activated eggs (described in Chapters 4 and 5), termed herein as **hPS cells** (human parthenote stem cells), and those obtained from egg cytoplasts following nuclear transplantation, termed herein as **hNTS cells** (human nuclear transplantation stem cells). The fundamental properties of each of these cell types have been described in preceding chapters. Neither of these types of human stem cells have been reported as of this writing.

hPS: *human parthenote stem cells.*

Primate parthenote stem cells (pPS cells) have been reported, however, and results to date are promising. A multiinstitutional research team that began with monkey egg activation by Jose Cibelli, working at Advanced Cell Technology, led to the development of four parthenogenically derived blastocysts from 23 mature monkey eggs. From the four blastocysts, one pPS cell line was derived. The cells displayed the same characteristics as other ES cell lines: they are chromosomally normal; appear capable of endless cycles of growth and cell division; contain significant amounts of the enzyme telomerase, which is thought to play a fundamentally important role in maintaining the integrity of chromosomes through cell division; and display antigens thought to be associated with ES cells. The pPS cells can be induced to differentiate into a variety of cell types in culture, including primitive neurons

HUMAN EMBRYONIC STEM CELLS

Immortal, pluripotential, karyotypically normal ES cells were isolated from mouse blastocysts more than 20 years ago. Most of the scientific community viewed them as a vehicle for genetic manipulations of mice, but a small group of scientists, mostly reproductive biologists, fully appreciated the scientific and therapeutic power of that breakthrough. One of those scientists was James A. Thomson at the University of Wisconsin. Another was Alan Trounson, a pioneer in human IVF and one of the founders of a premiere reproductive science center at Monash University in Melbourne, Australia.

James Thomson studied primate reproduction and had funding from the federal government to study primate ES (pES) cells. Once he had successfully isolated and partially characterized pFES cell lines from blastocyst ICMS, he knew similar work needed to be done to derive human embryonic stem cells. Confusion in federal funding agencies made it impossible to proceed in a scientifically logical manner to decide how best to derive hFES cells. Instead, Dr. Thomson decided to enter into proprietary agreements with Geron, a publicly held, for profit company. Using private capital, he equipped an entirely separate laboratory "across campus" from his federally funded laboratory so as to avoid expenditure of federal research funds on the derivation of hFES cells.

Alan Trounson pioneered many human reproduction methodologies, and began dreaming of isolating stem cells from human blastocysts at least a decade ago. His research efforts in Australia were also thwarted by strict government regulations about research with human embryos. He teamed with scientists in Singapore and Jerusalem to conduct his initial attempts.

Perhaps because "across campus" was logistically easier than the Australia-Singapore-Jerusalem connection, James Thompson was the first to report successful isolation and continuous culture of hFES cells in 1998. He not only characterized their chromosomal makeup, but also described the types of tissues that comprised the teratomas that resulted from transplantation of hFES cells to appropriate mouse hosts. Importantly, hFES cell lines appeared relatively easy to derive: from 14 ICMs, he derived 5 ES cell lines. Even before his work was reported in *Science*, the National Institutes of Health was seeking legal counsel on whether or not further research on the hFES cell lines could be funded with federal dollars. The outcome of that debate would take almost 3 years. Governor George Bush decided in 2001 that federal dollars could be used to study "existing" hFES cell lines, but that no federal dollars could be used to derive more hFES cell lines from human blastocysts.

Alan Trounson and collaborators published their successful isolation of hFES cells from the ICM of blastocysts in 2000. The characteristics of seemingly endless cycles of cell division and the capacity to differentiate into a wide variety of cell types were in agreement with the Wisconsin cell lines. Although this area of biomedical research remains highly controversial, the door to regenerative medicine has been opened.

capable of secreting dopamine, the brain cell signal missing in Parkinson's disease. The pluripotent nature of the pPS cells was further demonstrated in the teratomas formed following inoculation into appropriate mouse models. Tissues from all embryonic germ layers were formed, including blood vessels, primitive gut, smooth and striated muscle cells, bone, and cartilage.

These results suggest that, although parthenogenically activated primate eggs are not capable of supporting full development to an offspring, they may be capable of giving rise to pluripotent stem cell lines. In theory, hPS cells could be derived from the eggs of ovulating women with diseases amenable to stem cell therapy, such as Type I diabetes. At the time this book is being written, there are several unknowns with this approach.

First, although there is a high likelihood that hPS cells would be antigenically identical to and thus compatible with the egg donor (syngeneic), this must be proven because of the chromosomal crossing over events that occur during oocyte maturation. It is possible that the gene rearrangements could result in expression of proteins not normally expressed by the cells of the egg donor. Even if this proves to be the case, it is likely that the antigenic differences between the cells of an egg donor and hPS cells derived from her eggs will be far fewer than antigenic differences with hES cells from an embryo. Moreover, since each egg is likely to be unique, more than one hPS cell line could be screened for therapeutic suitability. Although the technology for deriving stem cells requires development of a blastocyst with an ICM, it is theoretically possible that stem cell lines in the future will be derived from individual blastomeres, which themselves could be screened for antigenic similarity to the egg donor.

Second, an homologous feeder cell layer from the egg donor needs to be developed to avoid cross-contamination of the hPS cells with potential mouse pathogens. This may or may not prove to be a significant problem. It is possible that skin fibroblasts, or cells from the ovary, may provide such a feeder layer.

The therapeutic potential of hPS cells is just beginning to be explored. This approach avoids not only all the ethical considerations associated with hFES cells or stem cells derived from fetal tissues, but may also eliminate the tissue rejection problems associated with allogeneic tissue transplantation.

Human Nuclear Transplantation Stem Cells

Analogous to animal cloning technologies, described in Chapter 10, the technical difficulties in deriving stem cells from egg cytoplasts into which nuclei from a donor has been transplanted will undoubtedly be solved in the near future. As of this writing, hNTS cells have not been described, although the work is ongoing in a number of laboratories. This technology will allow the development of therapeutic cell lines for older women and for men. As of this writing, one of the ethical stumbling blocks to federal funding for this research is the need to develop blastocysts with ICM from which to isolate stem cell lines. As described previously, if the laboratory conditions for deriving stem cell lines from cleaving blastomeres could be developed, not only would the process become more efficient, but the concerns of those in fear of cloning a human would be alleviated because a blastocyst would not develop. Since fertilization is not involved, and the goal is not to produce an embryo, but to produce a therapeutic somatic cell, a new term for this process is needed. One suggestion is ovasome, to describe the creation of somatic cells (-some) from eggs (ova-). Accuracy of language is especially important to this new field of biomedicine.

Ovasome: new term for using eggs to derive somatic stem cells.

The Process of Stem Cell Differentiation

To develop into fully functioning differentiated cells suitable for therapeutic purposes, pluripotent stem cells may need to go through the same stages of cell commitment as observed for stem cells in adult tissues. For example, blood stem cells first become committed to either lymphoid cell types or myeloid cell types (see Figure 1.1). As described in Chapter 1, these are referred to as progenitor cells, rather than stem cells, because it appears that once they have made the commitment, they do not revert to stem cells. In fact, the commitment process may be essential to preparing them for the next stage of differentiation. In addition, they may undergo a few cycles of cell division as progenitor cells. This may be an important mechanism for repopulating lost blood cells, especially in a crisis, such as hemorrhage. Studies of blood cell lineages from bone marrow stem cells are providing important information about the number and types of cell differentiation steps that are involved in the production of mature circulating blood cells from bone marrow stem cells. This information may provide clues about the essential steps involved in cell commitment and differentiation in other tissue types.

These considerations are reminiscent of the equivalence groups described in Chapter 8. Normal tissue and organ function depends on populations of cells that can respond with precision to specific signals, both from neighboring cells and from the body at large. Such responses are generally thought to be restricted to fully differentiated adult cells, not immature, progenitor cells. Given the utmost importance of balance between dividing cells and fully differentiated functioning cells that do not divide, it seems highly likely that tissues maintain pools of cells at defined stages of commitment, ready to respond when needed. Such pools of cells could be analogous to equivalence groups, positioned to respond in one way if called upon, or in another way if not called upon. The decisions could be time dependent. The nature of the response could be dependent on prior commitment events. Thus, for a mature lymphocyte to respond in a specific way, it may have had to progress through a defined pathway of differentiation events. An alternate, or deviant, pathway may result in a different response to the same stimulus.

Similar considerations may be relevant to stem cell differentiation for therapeutic purposes. For example, stem cell-derived neurons used to replace the brain cells that die in Parkinson's disease (described in Chapter 12) may need to proceed through a different series of commitment and differentiation steps than stem cell-derived neurons used to replace a damaged spinal cord. Understanding the specific steps is the challenge facing stem cell therapists.

Additional Readings

Jiang, Y., Jahagirdar, B., Reinhardt, R., Schwartz, R., Keene, C., Ortiz-Gonzales, R., et al. (2002). Pluripotency of mesenchymal stem cells derived from adult marrow. *Nature* 418: 41–49.

Kirschstein, R., and Skirboll, L. R., Eds. (2000). *Stem Cells: Scientific Progress and Future Research Directions.* www.nih.gov.

Marshak, D. R., Gardner, R. L., and Gottlieb, D. Eds. (2001). *Stem Cell Biology.* New York: Cold Spring Harbor Laboratory Press.

Niwa, H., Miyazaki, J., and Smith, A. G. (2002). Quantitative expression of Oct-3/4 defines differentiation, dedifferentiation or self-renewal of ES cells. *Nat Genet* 24: 328–330.

PART IV

Stem Cell Therapies

CHAPTER 12

Neurogenerative Diseases

The age-old idea that you can't make new neurons is being displaced by the notion that actually you can—if you know how to do it.

                                                           Ronald McKay

OVERVIEW

Until the 1990s, one of the central dogmas of neuroscience was that adult mammalian nerve cells could not regenerate. This was in contrast to amphibians and other lower animals that were known to regenerate nerves and even entire limbs. Not only could adult mammals not regenerate most nerve cells, they also could not repair damaged nerves.

Compelling evidence against this dogma was reported in the late 1990s by Elizabeth Gould at Princeton University. Dr. Gould took advantage of a relatively new technique for identifying cells in S phase by injecting animals with a modified DNA precursor molecule (termed BrdU) that mimics thymidine and is incorporated in its place during DNA replication. The methyl group on the thymidine was replaced with bromine, which could be identified by sensitive immunological techniques using antibodies developed specifically against bromine incorporated into DNA. The method also detects DNA being repaired, but the difference between S phase and DNA repair can usually be distinguished by the amount of bromine incorporated. In a series of experiments reported in the late 1990s, Dr. Gould demonstrated evidence of DNA synthesis in brain slices of rodents and primates.

In retrospect, two prior reports had also challenged the dogma that nerve cells cannot regenerate. In 1965, Joseph Altman and Gopal Das, working at

155

the Massachusetts Institute of Technology, published the results of studies that revealed fresh nerve growth in the adult rat hippocampus, but the reports were widely regarded as unusual exceptions to the dogma. However, in 1984, Evan Snyder, working at Children's Hospital in Boston, discovered that cells transfected with a reporter gene that readily stains blue had differentiated into neurons following transfer into a mouse. The neurons, which stained blue, had actually become integrated into brain tissue.

It was not until Elizabeth Gould reported her results that the accumulated evidence convinced Fred Gage of the Salk Institute that it was time to look in human brains for DNA synthesis in neurons. Although a seminally important question, there seemed no way to inject humans with the bromine-substituted DNA precursor to repeat the Gould experiments. Amazingly, Dr. Gage learned that a nearby cancer study included injecting terminally ill cancer patients with BrdU to evaluate tumor growth. He was able to obtain permission to study the brains of the cancer patients when they died. Brain sections revealed DNA synthesis in cells in the hippocampus (Figure 12.1), which was in agreement with Dr. Gould's rodent and primate findings.

Does this mean brain stem cells arise in the hippocampus? There is little information about the origin of brain stem cells, but work reported last year by Magdalena Gotz of the Max-Planck Institute of Neurobiology indicated that Pax6, a transcription factor expressed by neurons, could induce two supporting brain cells, radial glial cells and astrocytes, to differentiate into neurons. This is an important addition to the growing evidence that neural stem cells do exist. The challenge at hand is to learn to stimulate them to multiply and differentiate when needed to repair damaged nerve pathways.

In analogy to bone marrow cells, assume a "brain marrow" exists. This concept is supported by the growing body of evidence arising from work with embryonic stem cells. In the presence of both **epidermal growth factor (EGF)** and **fibroblast growth factor (FGF)**, ES cells can produce pluripotent **neurospheres**, which are aggregates of neural stem cells. Importantly, if EGF and FGF are removed, the neurospheres develop into cells with many of the characteristics of astrocytes. In contrast, if the cultures are supplemented with nerve growth factor (NGF), cells with the characteristics of neurons and

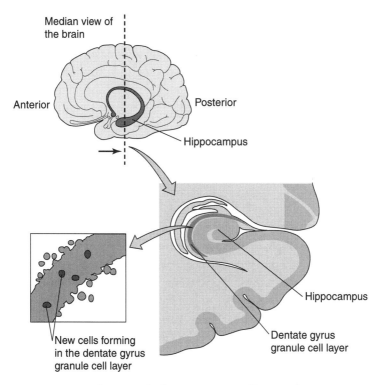

Figure 12.1 DNA synthesis in the hippocampus of human brain.

glial cells develop. Thus, the plasticity of nerve cells is beginning to be revealed.

Although neural fates are well mapped in lower animals, such as fruit flies and *C. elegans*, the sheer number of nerve cells and cell types in higher animals is staggering. Brain marrow is not as hectic as bone marrow, which pumps out tens of millions of new blood cells every second, but new brain neurons are formed each day at the rate of approximately 1 per 2,000 adult cells. It is clear that until much more is known about brain function, such as memory, it will not be possible to speculate how "brain cell turnover" could be accomplished. How, for example, could new neurons fit into the scheme of things? Are misplaced car keys the result of a new neuron learning the ropes?

Taken together, data gathered for the past decade show that the brain makes new cells when it learns, and it makes new cells to repair damage.

So, the idea of injecting cells into a damaged brain to speed up the process no longer seems as strange as it did a decade ago.

Parkinson's Disease

"At the moment the only source of unlimited numbers of the neurons appropriate for Parkinson's patients is ES cells."

Ron McKay

2000. *Nature* 406: 361–364

Parkinson's disease results from degeneration of dopamine-producing nerve cells in the substantia nigra (Latin for black substance), a tiny area of the midbrain that helps to control movement (Figure 12.2). Once these nerve cells die, they apparently do not grow back, at least not at a rate that alleviates the disease. The cells in the substantia nigra release a chemical, dopamine, which is one of a family of **neurotransmitters**, molecules that elicit a response from nerve cells. Substantia nigra neurons communicate with nerve cells in the **striatum**, which play a major role in controlling body movements. The result of loss of substantia nigra neurons is loss of muscle control, tremors, and stiffness. The disease usually affects people over 50 years of age, but it occasionally affects younger people as well, such as the actor Michael J. Fox. Former Attorney General Janet Reno and the athlete, Muhammad Ali, also suffer from Parkinson's disease.

> **Dopamine:** the neurotransmitter deficient in Parkinson's disease.

The cause of the loss of that specific type of brain cell is not known. Current treatment is focused on the administration of a chemical, **L-dopa**, which is converted in the brain to dopamine, and helps to make up for the loss of dopamine production by substantia nigra neurons. A principal problem with L-dopa administration is regulating the dose to meet the need. There is sometimes too much available, which causes involuntary muscle twitching and repetitive muscle movements, termed **dyskinesia**, and at other times too little available, which results in difficulty moving at all. Moreover, after a time, the body no longer responds to the L-dopa and disease symptoms worsen. In some forms of Parkinson's, memory becomes impaired and patients may become demented.

Replacement therapies with neuronal stem cells could be ideal for Parkinson's, since there is a specific location in the brain with known types of cells needing replacement. However, in order to be effective, the new neurons would need to also connect appropriately with the striatum. Unless the replacement neurons can establish themselves and integrate anatomically into the brain, the transplant would be unlikely to work.

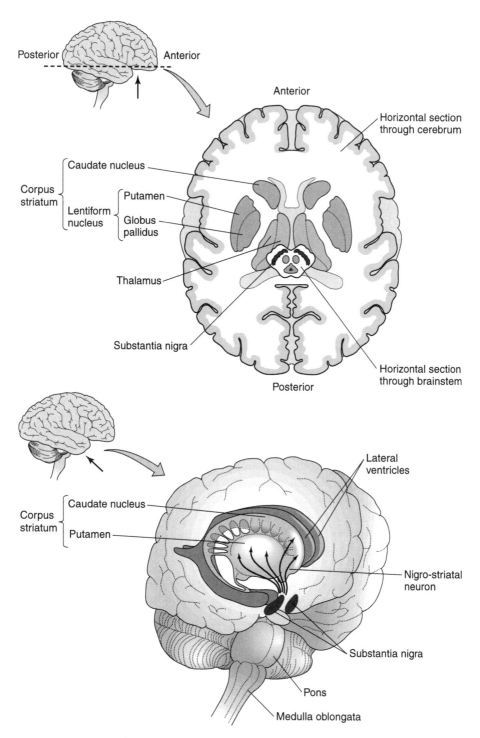

Figure 12.2 Area of brain affected by Parkinson's disease.

Studies transplanting adult nerve cells have never been shown to reestablish the appropriate connections. In the 1980s, however, researchers in Sweden studied the effects of transplanting fetal neurons from 7- to 9-week old human fetuses into the brains of victims of Parkinson's disease. Similar studies had previously been performed in mice and monkeys with some improvement in symptoms. Importantly, some alleviation of symptoms was observed in some of the patients, and autopsies conducted on a few who died from causes unrelated to either Parkinson's or the surgery revealed survival of and outgrowth of the grafted neurons, including into the striatum. Each procedure required tissue from several fetuses, however, and the danger of tissue rejection was thought to account for some of the failures. As discussed in Chapter 1, even though these stem cells are harvested from fetuses, they are nearly completely differentiated (see Chapter 9) and may not have the same developmental potential as stem cells.

Another similar trial recently conducted in the U.S. demonstrated no improvement in quality of life for the patients, but autopsies of two patients who died of unrelated causes showed that many of the dopamine neurons had survived and grown. Hope remains for this type of therapeutic approach to a disease that afflicts tens of thousands of Americans annually.

Cells that secrete and respond to dopamine (**dopaminergic**) are highly specialized and quite rare in the brain. Fewer than one cell in 100,000 brain neurons communicates through dopamine release. That rarity may relate to why they are so difficult to culture in the laboratory. The limited experiments with adult brain stem cells have shown that they produce differentiated dopamine-producing neurons for a limited period of time, and then stop. That has led researchers like Ron McKay at the National Institute of Neurological Disorders and Stroke, along with Lorenz Studer and Viviane Tabar at the Sloan-Kettering Cancer Center, to try ES cells. Unlike adult stem cells, ES cells can be cultured indefinitely, so they can be induced to form any number of dopamine-producing neurons desired.

As described in Chapter 8, the problem with ES cells is that they may form teratomas if they are injected while still undifferentiated. McKay and Studer have painstakingly developed procedures for culturing, differentiating, and purifying stem-cell lines. They have been able to make highly purified neuronal precursors from human ES cells that don't give rise to teratomas in immune-suppressed mice. In June 2000, McKay, along with Jong-Hoon Kim and Jonathan M. Auerbach, reported on their five-stage method of inducing ES cell cultures to differentiate into neurons (Figure 12.3). Using specific chemicals, the ES cells were first induced to differentiate into pluripotential neurons. Next, two types of neurons were developed— those that produce serotonin and those that produce dopamine. After the growth factor was removed from the cells, the neurons matured into fully functional nerve cells.

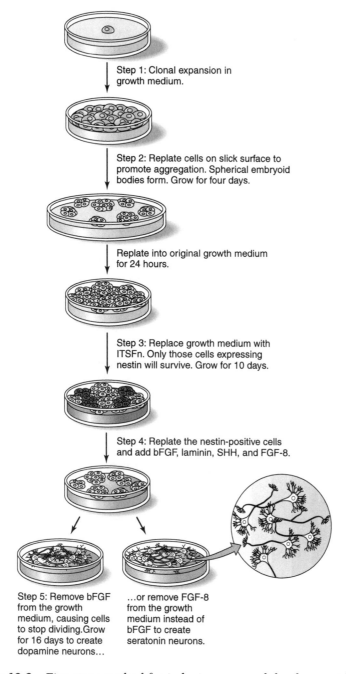

Figure 12.3 Five-stage method for inducing neuronal development from human ES cells.

In June 2002, they reported a transcription factor called Nuclear receptor related-1 (Nurr1) that plays a role in the differentiation of dopamine neurons. They reasoned that if they could cause that transcription factor to be expressed in the neurons developed from ES cells, they could enrich the population of precursors for dopamine neurons. They decided to use the technique of **transfection** to ensure expression of Nurr1 by some of the ES cells.

As described in Chapters 8 and 11, constructs of DNA can be taken up by cells in culture and become integrated into the cell's chromosomes. This process is termed transfection. The sidebar in Chapter 8 described the use of modified gene constructs whose ends were homologous to DNA sequences at either end of a gene sequence, thus allowing for targeted integration of new gene sequences, which replace the old gene sequences in the same location. Another approach, described in Chapter 11, takes advantage of other types of DNA constructs that include at either end the DNA sequences in some types of viruses (e.g., **retroviruses**) that facilitate integration into DNA, but not at specific gene sites. Thus, the integration of such constructs is more frequent than homologous recombination. The individual ES cells that take up the retroviral DNA constructs and integrate them into their genome are termed transgenic cell lines, and are analogous to the transgenic mice described in the sidebar in Chapter 5. Such constructs can also contain transcription promoter sequences that can be manipulated by the researchers, and thus the expression of the gene can be controlled by experimental conditions, rather than relying on the transcription factors produced within the cell itself. For example, a well-characterized transcription promoter sequence responds to the presence of steroid hormones. Thus, adding the hormone to the culture medium will lead to expression of the newly inserted gene. Because ES cells have the capacity to divide a seemingly endless number of times, transgenic ES cells can be individually selected and separated out into individual cultures. This process is termed clonal expansion (Chapter 11). McKay and co-workers selected ES cell clones transgenic for Nurr1 and subjected them to the five sequential stages of culture conditions that had previously been shown to generate dopamine-producing neurons (Figure 12.3). It worked.

They needed to test their cell lines in an animal model. Rats that have been chemically treated to kill dopamine neurons provide a good model for Parkinson's. After this treatment, the rats have impaired motion, analogous to a human Parkinson's patient. Using these rats, they grafted approximately half a million of the new neurons transgenic for Nurr1 into the rat's midbrain. The hope was that the mouse neurons would appropriately populate the human brain and not form a teratoma.

Amazingly, the mouse cells were accepted into the rat's brain and thrived. They displayed the electrical properties expected of midbrain neurons. They grew into the brain and out of the original area of the graft. McKay's team could even show that new synapses were being laid down, and that the mouse cells were "wired in" to the rat circuits. Most impressive was the rats' behavior. Rats with only half of their brain chemically lesioned favor the paw opposite the lesion. That was exactly what they saw in the control group. However, in the grafted rats, they noticed a remarkable improvement, close to what would be expected of normal, nonlesioned rats. Thus, based on behavioral tests, the rats were shown to have accepted the mouse cells, and were using them to offset the damage of their chemically induced Parkinson's.

The rats were examined for tumors and none were found, implying that the transplanted cells were free of undifferentiated ES cells. This experiment is vivid proof of the principle that ES cells can provide an unlimited source of controlled cells that can integrate into a host without causing tumors. To avoid tissue rejection, the ideal therapy would use stem cells derived from the patients themselves, such as through therapeutic cloning.

Alzheimer's Disease

"There is no treatment for Alzheimer's to reverse the disease or restore function once it is lost ..."

> *Ronald Petersen, M.D.*
> *Mayo Clinic*
> *Rochester, Minnesota*

Alzheimer's was first described by Dr. Alois Alzheimer in 1906. At autopsy, he noticed tangles and clumps called plaque in a patient with dementia. He described an ailment that started with memory loss and disorientation and led to language impairment, delusions, hallucinations, and death. The disease can take as little as 3 years or more than 20 years to run its course. Although many other diseases can cause dementia, Alzheimer's is by far the most common cause. Over a third of all people over age 85 have Alzheimer's. As that population group increases in number, Alzheimer's is expected to become a major medical issue around the world. Four million Americans have Alzheimer's today. That number is expected to triple in the next four decades.

The hippocampus, where memories are laid down, and the frontal lobes, the seat of logical thought, are affected first. Unlike Parkinson's, which affects neurons that communicate using dopamine, the affected nerves in Alzheimer's produce a different neurotransmitter, **acetylcholine**. These cells also send out long axons into distant brain structures. Early disease may be fairly localized, but as it progresses, Alzheimer's takes its toll in all areas of the brain. For stem cells to work, they will need to do more than merely take up residence in one area, such as the substantia nigra, in the case of Parkinson's disease. They will need to migrate to all the various structures of the brain that are damaged and integrate into the existing network of thickly connected neurons. Fortunately, this fits the characteristics of a neural stem cell. Apparently, in Alzheimer's patients, the normal repair system is just not fast enough to compensate for the ravages of the disease. Stem cells could provide the boost needed to balance the ongoing destruction, thus alleviating symptoms.

Recent research has found a genetic component to Alzheimer's. A specific allele of the apolipoprotein E gene predisposes a person toward Alzheimer's. That means if the stem cells could be engineered to eliminate this allele, such as through the homologous recombination described in Chapter 8, they might halt the onslaught and even reverse the damage.

Unlike Parkinson's, there have been no fetal cell trials in humans yet, but research with mice is quite encouraging. In April 2001, Kiminobu Sugaya, at the University of Illinois in Chicago, reported injecting aged, memory-impaired rats with neural stem cells. He found that the stem cells spread throughout their brains and that existing cells seemed to be rejuvenated. The rats were trained on a water maze, and the injected rats performed better than the controls. In fact, some of the elderly rats actually performed better than younger rats without memory impairment. One of the interesting aspects of this research is that human ES cells were used as a source for the transplants. This implies that ES cells may exhibit species-independent plasticity—finding a way to fit in, even in a different animal.

Other Brain Syndromes

"It is absolutely vital to continue research using embryonic neural stem cells. It may be that, for reasons we don't yet understand, adult stem cells will never be useful in therapy and that we will always need embryonic cells. Or, it may be the other way around. We just don't know."

Charles F. Stevens

Besides the neurons and the astrocytes, there is a third important cell type in the brain called the **oligodendrocytes**. These cells send out processes to the nerve axons and insulate them from the surrounding cytoplasm. The insulation is called **myelin** and without it, nerve impulses quickly damp out, slowing, or even halting communication (Figure 12.4). Diseases that involve a lack of myelin include multiple sclerosis, Krabbe's disease, and leukodystrophy. These diseases are systemic, and not localized to any particular cell type or brain location. Would stem cells be able to penetrate all the spaces necessary to myelinate axons?

Myelin: the insulating coating around neurons.

In 1999, Evan Snyder reported success using fetal human stem cells to treat knockout mice that had been bred to be deficient in myelin. These mice, called shiver or twitcher mice, are born normally, but develop tremors by the age of 2 weeks. Snyder injected the stem cells into the brains of newborn mice. The transplanted stem cells differentiated into oligodendrocytes and proceeded to wrap the naked axons with myelin.

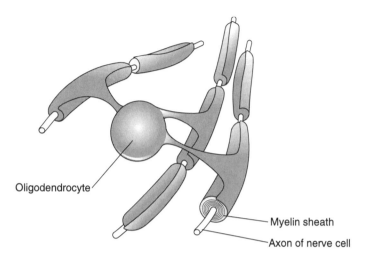

Figure 12.4 Myelin insulation to allow communication between nerve cells.

The cells migrated to all the parts of the brain that would be expected in normal development. In 60% of the mice, the tremors almost completely disappeared.

Snyder was able to expand his batch of human neural cells to quantities suitable for mice, but it still isn't clear that these adult stem cells can be proliferated indefinitely. Nevertheless, his research indicates that, given a suitable quantity of neural stem cells, some of these dreaded demyelination diseases may finally be treatable.

Ron McKay, along with Oliver Brüstle, a neuropathologist at the University of Bonn, has duplicated the Snyder research using neural stem cells derived from ES cells. McKay was looking for a plentiful supply of cells, and once again found ES cells to be the most reliable source.

Doug Kerr and Jeffrey Rothstein of Johns Hopkins are using stem cells with mice to develop treatments for ALS, also known as Lou Gehrig's disease. He used a virus called Sindbis, which destroys the lower motor neurons of the mice, causing them to drag their hindquarters when they walk. The neuronal death caused by the virus provides a good model for ALS, which is characterized by a similar death of motor neurons. In November 2000, he announced that experiments with stem cells were yielding positive results.

Rothstein and Kerr injected stem cells into the spines of the paralyzed mice and gave them time to migrate. He found that the stem cells migrated to the damaged area and that about 6% of them differentiated into neurons. Most importantly, half the mice recovered control of their hindquarters and were able to stand on all fours.

Using similar techniques, scientists hope to use stem cells to treat many other disorders of the brain. Animal trials are being conducted for Huntington's disease, Alzheimer's, strokes, Tay-Sachs, and alcohol impairment, with human trials expected soon.

Spinal Cord Injury

"Never before has there been such a powerful tool, such a resource that can give so much hope. And to have it just sitting here right in front of us, ready to go while all this debate rages on, is really, really frustrating."

Christopher Reeve

Approximately two million people worldwide are living with spinal cord injuries. One of the most famous is Christopher Reeve, who was paralyzed when he was thrown from a horse in an equestrian competition. Reeve has not only set up a foundation, but he is a powerful advocate for stem cell research.

The studies cited previously have demonstrated that new nerve cells can be encouraged to grow and integrate with a host brain. But many scientists were doubtful that they could reconnect a damaged spinal cord. It was also realized that more than just nerve cells were required. After damage, a cyst called a **syrinx** is formed and the nerves at the injury site become demyelinated, so oligodendrocytes are also needed to infiltrate and insulate the nerve cells.

In 1999, John W. McDonald, at the Washington University School of Medicine in St. Louis, reported positive results using ES cells on mice whose hindquarters were paralyzed from a spinal injury. Among other things, McDonald wanted to see what success he would get by waiting for a period before injecting the ES cells. Nine days after the injury, the mice were injected with neuronal precursors that had been derived from ES cells treated with **retinoic acid**. McDonald injected approximately a million cells into the syrinx that had formed around the site of the injury. The mice had been given cyclosporine to inhibit immune response and prevent rejection of the foreign tissue.

Retinoic acid: a molecule related to vitamin A that helps direct embryonic development.

A few weeks later, some of the mice were killed and their brains were scanned for the new cells. Although many of the cells had died, several had grown projections up to 1 centimeter long. Newly sprouted axons were clearly visible. After a month, several of the mice had recovered some of their mobility and coordination. Importantly, none of the mice developed tumors.

In December 2001, Hideyuki Okano at Keio University, Japan, announced that he had restored mobility to marmosets whose arms were paralyzed due to a spinal cord injury. All the marmosets, which are small monkeys, showed improvements after the transplant, and one regained almost full use of its arms. This marked the first success repairing spinal injuries in primates with stem cells. Okano used human fetal stem cells, which he was able to amplify to over a million cells, suitable for therapy in a small animal like a marmoset.

Since then, in August 2002, Okano was able to show that stem cells are more effective if they are infused a few days after the trauma. He noted that stem cells had a much better chance of differentiating into neurons if they were injected 9 days after the injury. That, coincidentally, is the same waiting period that McDonald used in his experiment.

Immediately following an injury, chemicals are released that inhibit neuron growth. After 9 days, these inhibitors have dissipated and neuronal growth can proceed.

Okano was able to obtain fetal neural stem cells from aborted fetuses. His technique involves culturing and expanding the fetal cells to get the greatest utility from them. Unlike ES cells, however, neural stem cells have not been proven to provide an immortal cell line.

Additional Readings

Altman, J., and Das, G. D. (1965). Autoradiographic and Histological Evidence of Postnatal Hippocampal Neurogenesis in Rats. *J Comp Neurol* 124: 319–333.

Eriksson, P. S., Perfilieva, E., Bjork-Eriksson, T., Alborn, A. M., Nordborg, C., Peterson, D. A., Gage, F. H. (1998). Neurogenesis in the Adult Human Hippocampus. *Nature Med* 4: 1313–1317.

Freed, C. R., Greene, P. E., Breeze, R. E., Tsai, W. Y., DuMouchel, W., Kao, R., Dillon, S., Winfield, H., Culver, S., Trojanowski, J. Q., Eidelberg, D., and Fahn, S. (2001). Transplantation of Embryonic Dopamine Neurons for Devere Parkinson's Disease. *N Engl J Med* 344: 710–719.

Jong-Hoon, K., Auerbach, J. M., Rodríguez-Gómez, J. A., Velasco, I., Gavin, G., Lumelsky, N., Lee, S-H., Nguyen, J., Rosario Sánchez-Pernaute, R., Bankiewicz, K., and McKay, R. (2002). Dopamine Neurons Derived from Embryonic Stem Cells Function in an Animal Model of Parkinson's Disease. *Nature AOP* published online 20 June 2002; doi:10.1038/nature00900.

McDonald, J. W., Xiao-Zhong, L., Qu, Y., Su, L., Mickey, S. K., Turestsky, D., Gottlieb, D. I., and Choi, D. (1999). Transplanted Embryonic Stem Cells Survive, Differentiate, and Promote Recovery in the Injured Rat Spinal Cord. *Nature Med* 5: 1410–1412.

Sang-Hun, L., Lumelsky, N., Studer, L., Auerbach, J. M., and McKay, R. D. (2000). Efficient Generation of Midbrain and Hindbrain Neurons from Mouse Embryonic Stem Cells. *Nature Biotechnol* 18: 675–679.

Wakayama, T., Tabar, V., Rodriguez, I., Perry, A. C. F., Studer, L., and Mombaerts, P. (2001). Differentiation of Embryonic Stem Cell Lines Generated from Adult Somatic Cells by Nuclear Transfer. *Science* 292: 740–743.

CHAPTER 13

Tissue Systems Failures

"ES cells… could provide a potentially unlimited source of islet stem cells that would be relatively straightforward to isolate and manipulate in culture."

Jon Odorico

OVERVIEW

Before any stem cell therapies can be considered for human trials, an important problem must be addressed. Currently, human embryonic stem cells are cultured on a thin mat of mouse fibroblast cells. To ensure that no animal products or viruses are communicated from this mat to the human ES cells, a new culturing protocol must be established. Fortunately, scientists in Singapore have managed to coax human cells into providing the proper growth environment. The technique, developed by Ariff Bongso and his team, at the National University of Singapore, also has the bonus of extending the culturing period from 7 to 9 days. This is the amount of time a cell culture can grow without spontaneously differentiating. To continue the line, cell cultures must be broken up and separated into different dishes. By extending the time between passages, a far greater number of cells can be cultivated between passages.

Singapore provides a peek into the future of stem cell research. Countries like Saudi Arabia, Israel, and Singapore, where the regulatory environment is favorable to stem cell science, will take the lead and determine the direction of research. Because President Bush has declared that federally funded research can only be performed on an existing set of embryonic cell lines—all of them grown on mouse fibroblasts—human clinical trials are unlikely to happen in this country until hPS cells or hNTS cells described in Chapter 11 are developed.

Diabetes

Diabetes affects over 170 million people worldwide, approximately 25 million of whom live in the United States. It is a disease characterized by the inability to either manufacture or use **insulin**, the protein hormone that regulates the uptake of glucose into cells. Insulin is secreted by specialized cells within the pancreas that are organized into regions named the islets of Langerhans.

> **Insulin: the protein hormone responsible for glucose uptake into cells.**

Without insulin, undigested sugars build up in the blood plasma instead of entering the glucose pathway (described in Chapter 6) within cells. Since glucose metabolism is a principal source of energy for human cells, the cells and the tissues they comprise are chronically energy deprived. Over time this causes widespread damage to tissues and organs.

There are two principal types of diabetes. Type 1 is also referred to as **juvenile-onset diabetes** and is characterized by undetectable levels of circulating insulin due to an absence of islet cells in the pancreas. Type 2 diabetes usually affects adults and is characterized by nearly normal levels of circulating insulin that does not stimulate normal glucose uptake by the cells. Type 2 diabetes accounts for at least 90% of U.S. cases, and can often be controlled by diet and exercise. One in five U.S. residents over the age of 65 has diabetes. Since the exact mechanism of Type 2 is still unknown, much work still needs to be done before a cure can be developed.

Type 1 Diabetes

On the order of one in two hundred U.S. children has Type 1 diabetes. Since Type 1 diabetes involves cell loss, it is a likely candidate for stem cell treatment. Stem cell therapy would involve differentiating stem cells into islet cells, or islet cell precursors, and then transplanting them into the patient. With an appropriate source and supply of islet cells, Type 1 diabetes should be possible to contain and even cure.

It has been shown that Type 1 diabetes can be cured with islets transplanted from donated organs in a procedure known as the Edmonton protocol. Unfortunately, the treatment calls for two pancreases per patient, and there are not nearly enough donated organs to go around. Only a few hundred of these procedures are possible each year, reaching only a small fraction of needy diabetics. Nevertheless, the success of the procedures provides proof that transplanted islets can integrate with the host tissue and function properly, actually curing this previously intractable disease. The successes immediately raised the need to isolate pancreatic stem cells and develop the methods to differentiate them into islet precursors to increase the supply of cells for therapy.

Unfortunately, pancreatic stem cells from adult organs have been frustratingly difficult to identify and isolate. Scientists do not yet know their morphology or what cell surface markers they display. Pancreatic stem cells also appear to have a fleeting existence in the developing fetus, so researchers are turning to embryonic stem cells, both to better understand the genesis of these cells and to obtain sufficient quantities for therapy. Nonetheless, whether the islet cells arise from adult tissue stem cells, fetal tissue stem cells, or embryonic stem cells, the immune system of the recipient must be suppressed to avoid rejection of the transplanted cells. Thus, a lifetime of injecting insulin is replaced by a lifetime regimen of immune-suppression drugs. Although the side effects from immune suppression are more controllable than the tissue pathology that results from Type 1 diabetes, it is clear that the ideal solution is to derive pancreatic stem cells from the diabetic's own cells.

Toward this end, several researchers have been developing protocols to induce the growth of pancreatic islet cells from ES cell cultures. In 2001, Bernat Soria, at the Universidad Miguel Hernández, was able to create insulin-secreting cells from mouse ES cells. Soria prepared a DNA construct that included a reporter gene (lacZ) and DNA sequences homologous to the flanking sequences of the insulin gene so that the DNA construct could be incorporated into chromosomes by homologous recombination, as described in Chapter 8. The insulin gene sequences were carefully chosen so that expression of the construct would be under the control of the promoter sequence that controls insulin production in islet cells. In this way, she hoped to be able to easily detect islet cell precursors at the first stage that they expressed insulin. Mouse ES cells were transfected with the DNA construct. The cells were then allowed to form embryoid bodies, with the hope that insulin-producing cells would differentiate out of the mass. It worked. Cells were selected for insulin production.

Soria injected the carefully selected cells into the spleens of mice with chemically induced diabetes. Within 24 hours, the mice were producing their own insulin and normalizing their glucose levels. Four weeks later, 60% of the mice were still normalized, indicating that the stem cells had successfully integrated into the bodies of the host mice. Twelve months later, some of these mice were still controlling their glucose levels, essentially cured of diabetes.

As promising as these results were, the ES cell procedure remains highly inefficient, yielding insulin-producing cells once for every 100 attempts. Unfortunately, the cells that grow well do not produce therapeutic levels of insulin, and those that produce insulin do not grow very well. Soria notes, "There appears to be an inverse relationship between the ability to proliferate and the ability to differentiate. We have not yet solved this problem, which again shows how much we need more research." Moreover, these studies also demonstrated that isolated cells do not produce the same smooth adjustments to glucose levels that are produced by islets that include three other cell types. This further complicates the production of appropriate islet cell replacements.

In February 2001, Jon Odorico and James Thomson, of the University of Wisconsin, demonstrated that primate ES cells can spontaneously differentiate into pancreatic cells. Odorico has also done work with human ES cells, but the work is progressing slowly. Odorico isolated a batch of cells that he believes are beta cells. They have been stained for analysis, but due to federal regulations, he cannot evaluate them with university microscopes. The results of his studies await private funding.

In April 2001, Ronald McKay and Nadya Lumelsky, of the National Institute of Neurological Disorders and Stroke, announced that they had been able to reliably differentiate insulin-producing cells from mouse ES cells. They allowed ES cells to form embryoid bodies and looked for those cells that expressed the protein **nestin**. Nestin is usually considered to be a neural cell inducer, but its presence is also associated with pancreatic islet cells. Using their five-stage culturing protocol (see Chapter 12), they induced the nestin-producing cells to assemble themselves into three-dimensional clusters that looked and acted like fully formed islets. When glucose was added to these structures, they produced insulin in response.

| **Nestin:** the protein expressed by progenitors to neurons.

These islets were then transplanted into the shoulders of diabetic mice, where they were quickly vascularized and supported. Unfortunately, the cells secreted a low level of insulin and were unable to totally cure the mice of diabetes. Nevertheless, this study showed that fully formed islets, composed of all four cell types and even including neurons, can be assembled out of ES cells. Note also that the locus of cell injection may not be important. Soria transplanted into the spleen and McKay's group simply chose the shoulder. Both locations tolerated the cells, vascularized them, and allowed them to produce insulin.

In August 2001, Suheirr Assady, at the Technion in Israel, showed that human ES cells could also be induced to form insulin-producing cells, taking another step on the path to human trials. Assady says, "These findings validate the human ES cell model system as a potential basis for enrichment of human beta cells or their precursors, as a possible future source for cell replacement therapy in diabetes."

Type 1 diabetes is thought to be an autoimmune disease in which islet cells are targeted for destruction by the victim's own immune system. If the diabetic's own islet cells were replenished, the risk is that the immune system might destroy them just as it did the original islets. However, it might be possible to genetically engineer these cell cultures so they won't be attacked by the immune system. If the genes that encode the cell surface proteins that target the islet cells for destruction could be excised, the islets might escape detection.

This theory was supported in 1993, when Peter G. Stock and his team at UCSF showed that islets from mice lacking the major class of cell surface proteins that usually stimulate immune rejection of transplanted tissues were not rejected. Therefore, a potential stem cell therapy might involve using a knockout vector to disable the MHC gene or genes. This "unmarked" ES cell line could then be cultured, cryopreserved, and stockpiled. When needed, the cells would be thawed and treated with special proteins to convert them to pancreatic stem cells. Once injected, these cells could repair the pancreas while avoiding the predations of the immune system, completely curing the disease with no side effects.

Alternatively, a therapy could be aimed at the misguided immune system itself. As in some cancer therapies, a technique may be developed to harvest a few stem cells, destroy the faulty immune system with radiation or chemotherapy, and then restart the immune system with the harvested cells. Research by Richard Burt suggests that once the immune system is renewed, it starts from scratch and no longer attacks the patient's own cells.

It may even be possible to modify autoimmunity with a simple hematopoietic stem cell transplant. Dr. Nagwa S. El-Badri and colleagues, at All Children's Hospital, University of South Florida, were able to completely halt rejection diseases in autoimmune-prone mice with a single allogeneic transplant. The mice, dubbed W/B F1, typically contract one of several autoimmune diseases by 8 weeks of age. However, after the transplant, 100% of the mice survived a year, completely free of autoimmune diseases.

Type 2 Diabetes

Type 2 diabetes afflicts ten times more people than Type 1 and is a major medical problem. Most people with Type 2 diabetes have insulin resistance, which is still poorly understood and may have multiple underlying causes. However, up to one quarter of Type 2 diabetics take insulin, so it may be possible to treat these cases like Type 1, by injecting islet cells derived from cultured ES cells.

If, as suspected, a genetic flaw lies behind Type 2 diabetes, then genetic engineering could repair the deviant gene. The corrected stem cells would be cultured and transplanted back into the patient. Since the stem cells are persistent and can spawn islet cells as needed, the diabetic might look forward to a complete cure with a single injection.

Rudolph Jaenisch and George Daley, from the Whitehead Institute for Biomedical Research, recently demonstrated the feasibility of this procedure using mice with a deadly genetic autoimmune defect. ES cells were created for each mouse using nuclear transfer. The flawed gene was then repaired in the lab. After culturing, these cells were injected back into the mouse. The transplanted stem cells were not rejected, formed stable cell populations, and partially cured the mice of their autoimmune disease.

Stem cells hold great promise for the treatment of the secondary diseases of diabetes, such as heart, kidney, eye, and circulatory ailments. As noted elsewhere in this book, each of these diseases is being targeted for research by stem cell scientists, who are hopeful for longer and healthier lives for all diabetics.

Cardiomyopathy

Cardiomyopathy, a disease of the heart characterized by weakened muscle fibers, affects over 4.8 million people in the United States alone. Most cases of cardiomyopathy are diagnosed at the time of a heart attack, which injures the heart and, due to its slow recovery time, leaves it vulnerable to further degradation.

Heart transplants are a last line of defense against a failing heart. There are far too few hearts available, and recipients must prepare for a continuing regimen of immune-suppressing drugs that leave them vulnerable to other infections.

It has long been assumed that cardiac muscle, like neural tissue, does not regenerate in adults. After a heart attack, the only observed changes seem to be cell death and the formation of scar tissue. However, new technologies have led to better observation, and it is generally now accepted that that some limited regeneration does indeed occur. **Cardiomyocytes**, the cells that comprise the pumping muscle of the heart, divide and start to repopulate the area. However, the response is often too little and too late. Unfortunately, no one has identified heart stem cells, which would presumably enhance the magnitude and speed of recovery. This is supported by the work described in Chapter 11 with mMAP cells.

| **Cardiomyocyte:** the precursor to heart muscle cells.

Nonetheless, other researchers have reported that hematopoietic stem cells derived from bone marrow can, in vivo, differentiate into apparent cardiomyocyte precursors. In 2001, Piero Anversa showed that injecting these marrow cells into damaged rat hearts would heal the tissue. The healing included heart muscle and the vasculature that supports the heart. Both modes of healing were evidently due to the fact that the injected cells were primitive enough to be able to differentiate into multiple tissue types.

In November 2000, Israeli researchers, headed by Nissim Benvenisty at the Hebrew University of Jerusalem, showed that human ES cells automatically generate cardiomyocyte precursors when they are allowed to form embryoid bodies. This work was extended by Lior Gepstein's group, at the Technion-Israel Institute of Technology, who showed more evidence that cardiomyocytes spontaneously appear in embryoid bodies. The scientists were careful to check expressed genes, transcription factors, enzymes, and electrochemical changes to verify that these cells were bona fide cardiomyocytes. Importantly, cardiomyocytes make themselves instantly known by their unmistakable rhythmic contractions. Without too much inducement, they simply differentiate out of mesoderm tissue and start pulsing. Gathered together by the millions and coordinated by nerve impulses, these tiny throbbing cells give rise to the impressive pumping action of the heart.

To repair a typical infarction, it may take a million or more cells. This argues against using adult stem cells, both because they are rare and because they are difficult to expand. Embryonic stem cells are better in this regard. Large quantities of cardiac precursor cells can be generated in vitro and then injected into the injured area. Because they are multipotent precursor cells, they can quickly expand their numbers in vivo, creating all the cell types needed for proper repair. To avoid rejection, the best cells would be derived from the patients themselves, using nuclear transfer. However, there is still an important issue that must be emphasized with regard to CS cells derived from ES clones: they are needed immediately after an attack. Longer waiting times lead to less positive outcomes, but the process of producing therapeutic quantities of cardiac stem cells after nuclear transfer can take weeks. To be most useful, then, patients would have to bank their own stem cells before an infarction.

In June 2000, the American Heart Association endorsed ES cell research and considered funding some studies. The AHA said that ES cell research represents "the most promising medical and scientific research" to fight cardiovascular disease. However, when they announced their intentions, a mail-in campaign spearheaded by Catholic Archbishop Justin Rigali of St. Louis convinced them to cancel their plans. As of 2002, there are still no federally funded heart studies that use human ES cells.

Kidney Failure

In June 2002, Robert Lanza, Michael West, and his colleagues at Advanced Cell Technology (ACT) reported that they had created a cow kidney from stem cells produced by nuclear transfer. After transplanting the kidney tissue under the skin of the donor cow, they found that it performed certain functions of a normal kidney, including the excretion of urine.

An important issue addressed by this study is the role of mitochondrial DNA in tissue rejection. Since nuclear transfer uses an enucleated egg, there is no nuclear DNA from the donor in the cloned cells. However, there are mitochondria in the egg that have their own DNA and replication machinery, and they do get passed on to the cloned stem cells. Might these egg mitochondria in some way produce an immune reaction? The study showed no rejection phenomena, thereby removing an important unknown about the efficacy of therapeutic cloning.

Liver Failure

Can liver tissue be repaired by stem cells? The evidence so far is positive. In November 2000, a team headed by D. S. Krause examined a number of liver biopsies from allografted bone marrow recipients. In particular, they were looking for women who had transplants from men and vice versa. They found two female and four male candidates. By staining for the X and Y chromosome, they were able to show that donor cells from the marrow had populated the livers of all subjects. The degree of infiltration was surprising—up to 43% of cells from some livers contained opposite-sex chromosomes. The largest infiltration was in a patient who had recurrent hepatitis C. This indicates that the infiltration was motivated by the injury site itself, and that bone marrow stem cells were recruited to differentiate into hepatic cells and repair the damage.

It is also known that ES cells can give rise to **hepatic** cells. Research done in 2002 by Hirobumi Teraoka showed that, after allowing ES cells to form embryoid bodies, hepatic cells start to show up within 12 days. These findings were supported by a separate Japanese study by Yukio Tsunoda that also showed that ES cells could form hepatic cells. The researchers found that hepatic cells spontaneously arose in embryoid bodies at around day 14. The hepatic cells took up an indocyanine green stain and formed distinct three-dimensional shapes in the embryoid body. These cells were found to produce the families of mRNAs that are characteristic of liver cells. Some of the cells were injected into the hepatic portal vein of other mice, where they integrated into the host liver and formed typical hepatic cells.

▌**Hepatic:** relating to the liver.

Cancer

As discussed in Chapter 1, cancer is a disease characterized by cells that no longer perform their intended function, and instead become highly proliferative and ultimately mobile and metastatic. In a sense, these cells are somewhat like primitive stem cells, without a declared function. Cells become cancerous on a regular basis. In general, because cancer cells usually express proteins on their surface that are not expressed by the noncancerous cells in the tissue involved, the cancer cells are discovered by the immune system and destroyed. If the cancer cell can hide or disguise its antigens, it may escape attack. Cancer cells may also avoid attack by amplifying a population of immune suppressor cells.

In a sense then, tumor growth and metastasis may be a malfunction of the immune system. And, in fact, bone marrow transplants that are prescribed for many cancers (see Chapter 1) aim to repair the immune system. The idea is to sensitize the harvested marrow to eliminate cancer cells.

The success of bone marrow transplants depends largely on the ability of stem cells to give rise to all the cells of the immune system. Recently, evidence has accumulated that marrow from closely, but not perfectly, matched donors may yield a better prognosis than autologous marrow (derived from the patient's own marrow). The speculation is that these cells do not have the same genetic deficiency as the patients' own cells. The problem is that the body's immune system may attack them in what is called host-vs-graft disease. Moreover, since a new immune system is being injected, there is the further problem of graft-vs-host disease, where the new immune cells target the patient.

The challenge is to find a donor graft that will target the cancer cells, but not the healthy cells. Despite the knowledge base about the genetics of immune rejection, achieving the proper balance is often difficult. Embryonic stem cells derived from the patient through nuclear transplantation or parthenogenetic stimulation of eggs, if the patient is a normally ovulating woman (described in Chapters 10 and 11), could eliminate a host of problems. Once ES cells have been derived from the patient, they could be amplified and induced to form blood-cell precursors, the primary regenerative component of bone marrow. Genetic deficiencies could theoretically be corrected in culture, before the cells are reinfused. In this way, patients could receive cells that are genetically identical to their own—with the exception of the corrected genes. These cells could both cure the cancer and avoid autoimmune destruction.

Hemophilia

Hemophilia is a genetic disease that prevents blood from clotting. Without the ability to clot, not only is death by exsanguination possible, but tissues have difficulty recovering from injury. About 17,000 Americans are afflicted by Type A or Type B hemophilia, which involve mutations in protein factors necessary to form a clot. Another bleeding disorder is von Willebrand's disease, which affects up to 2% of the population and involves a dysfunction of platelets. Because the disease is sex linked, hemophiliacs are almost exclusively men, but von Willebrand's disease affects both women and men. The long-standing treatment for hemophilia has been transfusions of clotting factors recovered from human blood donations. However, prior to the early 1980s, the blood supply was not being screened for human **immunodeficiency** virus (HIV), the causative agent of AIDS, and now an alarming 50% of hemophiliacs are infected by HIV. Of the original 10,000 men and boys infected in the early 1980s, 7,500 have died. Any treatment that can avoid the risk of exposure to pooled human blood products would be clearly superior. Stem cell therapies have been proposed.

Immunodeficiency: an inadequate immune response to pathogens.

Because a level of even 5% of normal clotting factor can greatly ameliorate the symptoms of hemophilia, the possibility of introducing blood stem cells that give rise to clotting factors merits study. In 2000, a Stanford study showed that hemophilia in dogs can be reversed by transfection of the blood coagulation factor IX (FIX) into skeletal muscle using a viral vector. The transecting agent, adeno-associated viral (AAV) vector, has also been used in murine studies. In mice that receive injections of the FIX/AAV complex, the levels of factor IX are substantially increased. These studies provide evidence that hemophilia can be cured with the proper expression of a single gene.

The same group in Roslin, Scotland, that produced Dolly the sheep also produced Poly, a sheep cloned from a transfected embryonic cell line. Poly produces Factor IX in her milk. Although Poly was engineered to produce therapeutic quantities of Factor IX in her milk, it also demonstrates that nuclear transfer can be combined with genetic manipulation to create persistent genetic changes in mammals other than cows, as described in Chapter 10.

The lesson of these studies is that many genetic diseases, especially those with a well-known etiology and a single genetic defect, can be addressed with embryonic stem cell technology. Because the stem cells have the ability to be passaged indefinitely, they are reasonable targets for genetic manipulation of the sort that produced Poly. Other diseases that fall into this category include familial hypercholesterolemia, severe combined immunodeficiency (SCID), cystic fibrosis, and certain cancers.

Additional Readings

Assady, S., Maor, G., Amit, M., Itskovitz-Eldor, J., Skorecki, K. L., and Tzukerman, M. (2001). Insulin Production by Human Embryonic Stem Cells. *Diabetes* 50: 1691–1697.

Chinzei, R., Tanaka, Y., Shimizu-Saito, K., Hara, Y., Kakinuma, S., Watanabe, M., Teramoto, K., Ariil, S., Takase, K., Sato, C., Teradae, N., and Teraokae, H. (2002). Embryoid-Body Cells Derived from a Mouse Stem Cell Line Show Differentiation into Functional Hepatocytes. *Hepatology* 36: 22–29.

Itskovitz-Eldor, J., Schuldiner, M., Karsenti, D., Eden, A., Yanuka, O., Amit, M., Soreq, H., and Benvenisty, N. (2000). Differentiation of Human Embryonic Stem Cells into Embryoid Bodies Comprising the Three Embryonic Germ Layers. *Mol Med* 6: 88–95.

Jacobson, L., Kahan, B., Djamali, A., Thomson, J., and Odorico, J. S. (2001). Differentiation of Endoderm Derivatives, Pancreas and Intestine, from *Rhesus* Embryonic Stem Cells. *Transplant Proc* 33: 674.

Kay, M. A., Manno, C. S., Ragni, M. V., et al. (2000). Evidence for Gene Transfer and Expression of Factor IX in Haemophilia B Patients Treated with an AAV Vector. *Nat Genet* 24: 257–261.

Kehat, I., Kenyagin-Karsenti, D., Druckmann, M., Segev, H., Amit, M., Gepstein, A., Livne, E., Binah, O., Itskovitz-Eldor, J., and Gepstein, L. (2001). Human Embryonic Stem Cells Can Differentiate into Myocytes with Structural and Functional Properties of Cardiomyocytes. *J Clin Invest* 108: 407–414.

Lanza, R. P., Chung, H. Y., Yoo, J. J., et al. (2002). Generation of Histocompatible Tissues Using Nuclear Transplantation. *Nat Biotechnol* 20: 689–696.

Lumelsky, N., Blondel, O., Laeng, P., Velasco, I., Ravin, R., and McKay, R. (2001). Differentiation of Embryonic Stem Cells to Insulin-Secreting Structures Similar to Pancreatic Islets. *Science* 292: 1389–1394.

Nathwani, A. C., Davidoff, A. M., Hanawa, H., Hu, Y., Hoffer, F. A., Nikanorov, A., Slaughter, C., Ng, C. Y. C., Zhou, J., Lozier, J. A., Mandrell, T. D., Vanin, E. F., and Nienhuis, A. W. (2002). Sustained high-level expression of human factor IX (hFIX) after liver-targeted delivery of recombinant adeno-associated virus encoding the *hFIX* gene in rhesus macaques. *Blood* 100: 1662–1669.

Orlic, D., Kajstura, J., Chimenti, S., Jakoniuk, I., Anderson, S. M., Li, B., Pickel, J., McKay, R., Nadal-Ginard, B., Bodine, D. M., Leri, A., and Anversa, P. (2001). Bone Marrow Cells Regenerate Infarcted Myocardium. *Nature* 410: 701–705.

Osorio, R. W., Ascher, N. L., Jaenisch, R., Freise, C. E., Roberts, J. P., and Stock, P. G. (1993). Major Histocompatibility Complex Class I Deficiency Prolongs Islet Allograft Survival. *Diabetes* 42: 1520–1527.

Richards, M., Fong, C. Y., Chan, W. K., Wong, P. C., Bongso, A. (2002). Human Feeders Support Prolonged Undifferentiated Growth of Human Inner Cell Masses and Embryonic Stem Cells. *Nat Biotechnol* 20: 933–936.

Soria, B., et al. (2000). Insulin-Secreting Cells Derived from Embryonic Stem Cells Normalize Glycemia in Streptozotocin-Induced Diabetic Mice. *Diabetes* 49:157–162.

Theise, N. D., Nimmakayalu, M., Gardner, R., Illei, P. B., Morgan, G., Teperman, L., Henegariu, O., and Krause, D. S. (2000). Liver from Bone Marrow in Humans. *Hepatology* 32: 11–16.

Yamada, T., Yoshikawa, M., Kanda, S., Kato, Y., Nakajima, Y., Ishizaka, S., and Tsunoda, Y. (2002). In Vitro Differentiation of Embryonic Stem Cells into Hepatocyte-Like Cells Identified by Cellular Uptake of Indocyanine Green. *Stem Cells* 20: 146–154.

Yin, Y., Lim, Y. K., Salto-Tellez, M., Ng, S. C., Lin, C-S., and Lim, S-K. (2002). AFP+, ESC-Derived Cells Engraft and Differentiate into Hepatocytes in Vivo. *Stem Cells* 20: 338–346.

PART V

Human Embryonic Stem Cells and Society

CHAPTER 14

The Religious, Legal, Ethical, and Scientific Debate

Despite many claims to the contrary, life does not begin at conception: It is an unbroken chain that stretches back nearly to the origin of the Earth, 4.6 billion years ago.

Carl Sagan, 1998

OVERVIEW

As discussed in Chapter 11, Dr. James Thomson used surplus blastocysts from a fertility clinic for his stem cell research. The embryos, scheduled for destruction, were donated by the parents for this project. He knew the federal funding for his animal research at the University of Wisconsin could be withdrawn if it overlapped in any way with his privately funded human embryo research. He had taken great pains to separate the labs, not sharing so much as an extension cord between the two. It was expensive and redundant, but absolutely essential because the federal law was unambiguous in its ban on funding human embryo research, first stated in 1996, and renewed in 1998. This ban had originaly created an unprecedented void in the United States of basic science research to underly the clinical treatment of infertility by assisted reproduction, and it now extended to embryonic stem cell research.

Related to the federal ban on research was the underlying view held by many U.S. citizens that life begins at fertilization, and thus research on fertilized eggs was likened to an early abortion. There is no doubt that elective abortion is a tragedy for all involved. Some antiabortion groups have become extremists and feel justified in killing doctors who perform abortions. They also maintain Internet hit lists that provide names and addresses of doctors and scientists who work in any capacity with human embryos. The irony is that by stopping research on fertilized eggs and early embryonic development, the extremists actually inhibit the scientific understanding of the process, and thus the development of new ways to prevent pregnancy, and in this way perpetuate the need for abortion itself. There is no substitute for knowledge.

Thus, to Jamie Thomson's concerns about federal funding for his nonhuman research were added the additional worries about being on the hit list of radical groups. On the one hand, concern for each and every human life speaks well for the values in our society. On the other hand, when should the concern begin and who should establish the ethical guidelines? These are complicated social issues that will require much public debate to resolve and are based on centuries of religious and theoretical dogmas.

The Religious Debate

The ablation of the inner cell mass (ICM) of the blastocyst, which critically and irremediably damages the human embryo, curtailing its development, is a gravely immoral act and consequently is gravely illicit.

Vatican City
August 25, 2000

Some, but by no means all, religions believe that human life starts at conception. The Roman Catholics are the most powerful and adamant about this concept, but they weren't always so. In the fifth century, Saint Augustine proclaimed that abortion before the **quickening**—when the baby becomes large enough to have its movements felt by the mother, approximately 4 to 5 weeks of development (Chapter 9) —was neither a sin nor a homicide, because "one cannot say that there is a living soul in a body that lacks sensation ... the flesh has not yet been formed and thus has no capacity to feel."

Quickening: the sensation in the uterus caused by movement of the fetus.

In the twelfth century, Saint Thomas Aquinas advanced the theory of delayed "hominization," or "ensoulment." Aquinas saw embryogenesis as proceeding through three stages: vegetative, animal, and rational. Only in the ultimate, rational stage did the fetus gain a soul and therefore become a human being. As did his predecessors, Aquinas pegged that event at day 40 (counting from the last menstrual period), concluding that abortion before then could not be considered murder.

At the start of the thirteenth century, Pope Innocent III reiterated that the quickening was still considered to be the start of life and that prior to day 40, abortion was not homicide. From 1140 to 1917, that was canon law. In the sixteenth century, Pope Gregory XIV confirmed once again that life starts at the quickening, and that abortion before then was not homicide. In 1869, however, Pope Pius IX changed the direction of the church. For the first time in Catholic history, abortion—from the moment of conception—became the equivalent of homicide. The 1917 Code of Canon Law put it all in writing.

What convinced the church that conception was the magic moment when life begins? It was the explanation by scientists that fertilization involved sperm and eggs. Observations since van Leeuwenhoek first spotted sperm with his microscope led to the establishment of embryology in the 1800s. By the middle of the century, it was established that sperm and eggs were both necessary and sufficient for the formation of a zygote, although it was still unclear what parts of these cells contained the hereditary components.

Thus, for nearly 2,000 years, the church accepted abortion in one form or another, and only for the last 150 years has it been banned, based on the science of that era. As science once again shifts paradigms, the Roman Catholic church may again reinterpret its canon.

If the history of the Roman Catholic religion is not a picture of unwavering conviction on the issue of abortion, neither is any other religion. Religious thinkers from all major religions have come down on one side or another, at one time or another.

For example, Islamic scholars are not unanimous on abortion—and some imams strictly prohibit it—but the general theory is similar to Aquinas' delayed ensoulment. Islam usually accepts abortion before the 80th day, but also believes that the sin increases with the duration of the pregnancy.

Buddhists treat life as a continuum, both in time and between species, so it is possible to consider a bug to be as sacred as an embryo. Abortion is therefore wrong and negatively affects your karma. However, there are rituals to dispel bad karma, so the stain of sin may be temporary. Most modern Buddhists agree that the decision of whether or not to abort belongs to the mother.

Judaism, like Islam, is not unified on the issue of elective abortion, but in general, the consensus among Reformed Jews is that abortion is a woman's choice. Although the fetus is still sacrosanct, it isn't considered a person until it takes its first breath of life at birth—the Judaic version of delayed ensoulment.

The Presbyterian church supports a woman's right to choose, as do many other American Christian groups.

In summary, although some religions have strong convictions, the opinion of religious scholars is prismatic and fractured. Although the majority of thinking comes down on the side of a mother's right to choose, there is plenty of dissension. Abortion is an issue that affects men and women differently, and abortion rights often track advances in women's rights. As liberal governments have flourished, human rights have expanded, and today about two thirds of the world's population lives in countries where women have at least nominal rights and abortion is legal.

Religious opinions of embryonic stem cell research closely track positions on abortion. The religions that are liberal on abortion are generally positive about stem cell research.

For example, The Church of Scotland "affirms the special status [of the embryo] as created by God, but it recognizes the potential benefits of embryo research under limited circumstances." It approves studies with nuclear transfer as well, but hopes that its use will be limited and aimed at ultimately creating therapies that will remodel somatic cells without using an egg.

Bert Clifford, the chair of the Ethics Committee at Sydney Adventist Hospital, says a ban on stem cell research has no moral justification. "We should be open to dialogue, open to discussion, open to possibilities, open to finding the best solution in changing times in advancing technology."

The Episcopalian Church is divided, but some bishops claim that embryonic research should be permissible on embryos up to 14 days old.

The Presbyterian stance is also favorable and includes the donation of embryos from fertility clinics. The church has issued a statement saying that "Prohibition of the derivation of stem cells from embryos would elevate the showing of respect to human embryos above that of helping persons whose pain and suffering might be alleviated. Embryos resulting from infertility treatment to be used for such research must be limited to those embryos that do not have a chance of growing into personhood because the woman has decided to discontinue further treatments and they are not available for donation to another woman for personal or medical reasons."

The Union of Orthodox Jewish Congregations expressed similar sentiments in 2001, saying, "an isolated fertilized egg does not enjoy the full status of personhood."

At the other end of the spectrum, the Lutheran and United Methodist churches agree with the Vatican, and are opposed to human embryonic stem cell research. Similarly, the Southern Baptist Convention asserts, "Human embryos are the tiniest of human beings."

Another impassioned voice against human embryonic stem cell research is the American Life League (ALL), which grounds its theocracy in Roman Catholic teachings. ALL bills itself as the largest grass-roots, pro-life organization in the United States. They want to criminalize embryonic stem cell research, and they also want to ban fertility clinics and contraceptives.

The erratic history of religious pondering on the topic of when life begins indicates that the topic should not be considered closed. The forgiveness extended to both Galileo and Darwin by the Roman Catholic church demonstrates that the Roman Curia can change its mind and accept scientific evidence. Although the Pope may be considered to be infallible, no such claim is made for Catholic law, which is allowed, and expected, to change.

Science has confirmed Saint Augustine's observations that the "flesh" of an early embryo is indeed without form and bereft of sensation. Perhaps the most curious aspect is that, despite the high esteem held by the church for both Aquinas and Augustine, and the fact that modern science seems to be vindicating their observations, the opposite conclusion is being currently propounded by the Vatican.

The religious argument against ES cell research would be much stronger if all religions agreed. Instead, not only do many religions find the research to be permissible, but some even believe it would be immoral *not* to use stem cell therapies. The problem is difficult and emotions run high. Religions are attempting to reconcile the overlapping rights of three parties: the mother, an embryo, and the patient. The answer is not an easy one.

The Legal Debate

The Department of Health and Human Services (DHHS) concluded that the Congressional prohibition [on the funding of embryo research] does not prohibit the funding of research utilizing human pluripotent stem cells because such cells are not embryos.
National Institutes of Health Guidelines for Research Involving Human Pluripotent Stem Cells, 2001

In many countries, the legal disposition of abortion is dictated by religious leaders. Most western countries, however, have attempted to separate religion and law. Modern societies, where travel and foreign trade flourish, find it difficult to impose any single religion on diverse citizenry. Secular civil law, based on democratically elected governments, has proven to be a successful formula for economic and scientific progress. Unfortunately, leaving religion out of the legal system has not really simplified matters. The legal standing of an embryo has been in constant flux throughout American history. It has variously been defined as a human, as property and, more recently, as a new type of entity that isn't owned, but whose disposition belongs to the gamete donors.

In the 1930s, researchers routinely used fetal tissue from abortions and miscarriages for experimentation and, most importantly, as an invaluable source of cultured cells for studying bacteria and viruses. In 1949, it was a human fetal kidney cell culture that finally allowed Harvard researcher John Enders to grow the polio virus in vitro. This work ultimately led to the development of the polio vaccine by Jonas Salk and Albert Sabin. For this life-saving breakthrough, Enders was awarded the 1954 Nobel Prize for Medicine.

In 1973, ignoring the strongly stated objection of the Roman Catholic church, the United States Supreme Court ruled in Roe v. Wade that a woman has a legal right to terminate her pregnancy up until the point where the embryo is viable and able to survive on its own outside the womb. The court concluded that a fetus is not a person within the meaning of the Fourteenth Amendment. Thus, the idea that a 5-day old blastocyst is a human being is not supported by U.S. law.

Nonetheless, the emotions inflamed by the Roe v. Wade decision led to a moratorium on federally funded embryo research. The following year, Congress established the National Commission for the Protection of Human Subjects of Biomedical and Behavioral Research (**CPHSBBR**), which sought to regulate embryo research.

CPHSBBR: *Commission for the Protection of Human Subjects of Biomedical and Behavioral Research.*

In 1987, the NIH placed a moratorium on federal funding of all human embryo research until an advisory panel could issue guidance on the types of research that should be permitted. In 1988, the panel voted 19–2 in favor of funding for human embryo research. But, this recommendation not withstanding, Louis Sullivan, the Secretary of the DHHS at the time, sided with the two dissenters and continued the moratorium. Subsequently, the U.S. Congress twice voted to override the DHHS moratorium, but then President George Bush the elder vetoed the votes. The ban on federal funding stayed in effect.

In 1993, Donna Shalala, the director of the DHHS during President Clinton's administration, lifted the moratorium and Congress passed the NIH Revitalization Act. This act provided a regulatory framework for research that stipulated that any embryonic tissue could be used, but only for therapeutic purposes. The tissue would have to be accompanied by written permission from the mother or donor affirming that the donation was for therapy, that no restrictions would be placed on the recipient, and that the recipient would remain anonymous. The act also required that the mother make the decision to abort before and separately from the decision to donate tissue. This was to ensure that the option of donating embryos for research did not encourage women to choose abortion. There is, however, no known correlation between embryo research and abortion, anymore than there is between using the organs of murder victims and the rate of murder.

These regulations govern federally funded research, but they do not carry criminal sanctions. There are, however, some explicit laws governing what can be done with embryos and/or fetal tissue. It is against the law to profit from interstate commerce in embryos, it is illegal to solicit or accept donated tissue when the recipient is a friend or a relative, and it is illegal for the recipient to pay for the procedure. In all cases, the ultimate arbiters of the disposition of the fetus are the parents.

In 1994, another panel convened to advise Congress on the issue of human embryo research again voted in favor of research. At first, President Bill Clinton promised to agree with the panel's decision in the form of an executive order, but after a well-organized write-in campaign by Christian groups, including the Roman Catholic College of Cardinals, then-President Clinton decided not to set aside the ban on federal funding.

It is important to note that, although federal funding for embryo research is unavailable from the government, the research itself is not illegal. This distinction has been lost on many in the anti-stem cell movement, including Senator Sam Brownback, who has written that "embryonic stem-cell research is illegal, immoral and unnecessary." In fact, it was and still is completely legal to do embryo research with private financing. It was private funding from Geron that allowed James Thomson to isolate the first human ES cells.

President Bush's declaration, on August 9, 2001, that he would allow stem cell research on 64 existing embryonic cell lines, opened the lid slightly on federal funding. He stated, "I have concluded that we should allow federal funds to be used for research on these existing stem cell lines where the life and death decision has already been made." Unfortunately, these cell lines—for reasons relating to the infancy of the research—are of uncertain utility. Some are not expanding properly, and others seem to have lost their capacity to differentiate. All of the lines have apparently been grown on mouse fibroblasts, making them unsuitable for human therapy. Comparing data on these cell lines is complicated by the fact that these cells were created in independent research labs around the world, each following separate protocols.

However, that decision is rightly the parent's to make, not the government's. There are hundreds of thousands of frozen embryos left over from fertility procedures around the world that could be used to create new embryonic stem cell lines using better, standardized procedures. These cell lines, like blood and bone marrow, are most useful when there is a tissue type match. That is a powerful argument in favor of more embryonic stem cell lines, not just the limited set proposed by the President.

The number that have been made available, along with the red tape that it takes to procure them, compounded further by conflicting patent rights and ownership claims, has resulted in very few federally funded embryonic stem cell studies. As of October 2002, only four stem cell material transfer agreements had been signed.

As disappointing as these results are, they nevertheless point to a loosening of the government's ban on embryo research, hastening the promise of eventual therapy. However, before many of the NIH grants go out, the Congress may ban certain aspects of this research entirely. On August 1, 2001, the House of Representatives voted overwhelmingly (265–162) to make nuclear transfer using human cells a crime, punishable by a million dollar fine and 10 years in jail. This will become the law of the land if the Senate version of the bill (SB-790) is also passed. Denying federal funding has set back embryo research for over a decade. The sponsors of the ban hope to put such a chill into the field that they can stop it entirely, forcing such research to move overseas.

That is happening already. Several researchers have left the United States to start new research labs in England, Singapore, Israel, Saudi Arabia, and Japan, where the laws are more liberal and the governments are promoting the science. Each of these countries expects to benefit greatly from American intransigence.

Moving in the opposite direction, in September 2002, California governor Gray Davis signed a law explicitly promoting stem cell research, including nuclear transfer. California may even provide funds for human embryonic stem cell research. This law supersedes President Bush's Executive Order, but it, in turn, could be superseded by Federal law, should the Senate pass SB-790.

The Ethical Debate

[Researchers] should be asking who this research should be done on, will it work, what will the cost be and who should have access to it. But because of abortion politics, we spend no time at all talking about this.

Arthur Caplan
University of Pennsylvania Center for Bioethics

The ethical debate is in the eye of the storm about human embryonic stem cells, and those attempting to make calm judgments do so as the windy arguments between religion, the law, and science churn around them. Bioethics committees are an important part of any research project. Most scientists welcome their recommendations, as it lets them proceed with a minimum of moral ambiguity.

The first bioethicist of record was Hippocrates, the author of the injunction to "first, do no harm." For over 2,000 years, his paternalistic vision prevailed, and helped medicine to become a professional, almost priestly trade. Doctors, it was believed, could be counted on to produce wise and moral judgments about a patient's health—judgments that were considered too difficult for the patient to possibly comprehend. But this comfortable relationship between the childlike patient and the authoritative doctor-minister was not to last.

The end of World War II exposed the depredations of Nazi doctors who performed experiments on unwilling human subjects. In response to these horrors, the

Nuremberg Code was created to establish a list of human rights that must be observed by all moral and ethical research. These principles are summarized as follows:

1. The voluntary consent of the human subject is essential.
2. The experiment should yield fruitful results for the good of society, unprocurable by other means.
3. The experiment should be based on the results of animal experimentation and the anticipated results should justify the experiment.
4. The experiment should avoid all unnecessary physical and mental suffering and injury.
5. No experiment should be conducted where there is a reason to believe that death or disabling injury will occur.
6. The risk to be taken should not exceed the problem addressed by the experiment.
7. Proper preparations should be made to protect the experimental subject.
8. The experiment should be conducted only by scientifically qualified persons.
9. The human subject should be at liberty to end the experiment.
10. The scientist must be prepared to terminate the experiment if he has cause to believe that the experiment is likely to result in injury to the experimental subject.

For persons who believe that blastocysts are human beings, stem cell research is clearly unethical on these grounds. For persons who do not believe a blastocyst is a human, the second clause, which requires the consideration of other means of achieving the same goals, such as adult stem cells, remains problematic. Otherwise, stem cell research satisfies these guidelines.

The Nuremberg Code was, unfortunately, not enough to stop the infamous Tuskegee Syphilis Study in Alabama. This research project, approved and reapproved over a 40-year period by the U.S. Health Services, denied poor black men treatment for syphilis, even though the cure (penicillin) was discovered during the trial. The Tuskegee project not only brought well-deserved opprobrium to the Health Services, and much stricter rules about informed consent, but proved that federal institutions should not be considered the ultimate authority on ethical guidelines for research.

IRB: *internal review board*, the institutional committee that enforces guidelines for research involving human subjects.

Bioethics committees now abound. Most universities have an **Institutional Review Board (IRB)** or a Research Ethics Committee (REC) to clarify the legal and ethical issues involved with any kind of human research. Private companies have set up bioethics committees composed of legal, religious, and ethical scholars. Capping it all off is the National Bioethics Advisory Commission (NBAC) established by President Clinton in 1994 and continued by President George Bush.

Not surprisingly, ethicists are about as unified as religious scholars on the merits and dangers of ES cell research. However, their view of a human almost always includes sentience or at least a history of sentience. On that basis, most ethicists don't consider a blastocyst to be a person. An argument can also be made that the blastocyst is not destroyed, since the cells that were destined to become the embryo are exactly the stem cells that are cultured. Since these cells are still alive, and since they are still in an undifferentiated state, in what sense has the blastocyst been killed? One can't even say it is "dismembered," since there are no members in a blastocyst. However, if full personhood is accorded to a blastocyst, it would still be objectionable to disassociate the cells, even if death were not the result. Nevertheless, the notion that a blastocyst could be disassembled and then reassembled puts it into a special ethical category distinct from normal human beings.

An ethos that protects the human right to privacy could view ES cell therapy as a simple private arrangement between, say, spouses who wish to fertilize an egg to create stem cells for their own therapy. If spouses are granted the privacy to make children and they are not required to report miscarriages to the authorities, then would it be ethical to create a blastocyst that might have therapeutic qualities? If this level of privacy is not a right, and if blastocysts are human beings, then all pregnancies would need to be reported to some legal agency that would track them to assure that each one comes to term and is not diverted for therapeutic purposes.

A simpler ethical decision is whether or not to use surplus embryos beyond the ones already specified by President Bush. There are hundreds of thousands of frozen blastocysts in fertility clinics around the world and the vast majority of them are destined to be destroyed. The simple ethical conclusion is to use them. In fact, most scholars believe that it is unethical to let sick people die by denying them a potential treatment. That is, after all, what happened in the Tuskegee study.

The argument against using the surplus blastocysts is that it might encourage people to create embryos for therapy. However, the law prevents people from profiting from the sale of embryos, so the only reason to create a therapeutic embryo would be compassion. However, since some Christian groups do not call it compassion, but abortion, it is the job of the ethicist to try to reconcile these two diametrically opposed views.

A starting point for many ethicists is the utilitarianism of Jeremy Bentham and John Stuart Mill, who argued that the best civilizations will perform "those acts which result in the greatest good for the greatest number of people." The point is to maximize happiness and minimize pain for all involved. If those involved include an insensate blastocyst, not yet capable of feeling either happiness or pain, and a sick patient who could benefit from that tissue, then the conclusion is easy: Sacrifice the blastocyst for the good of the sick patient. To do otherwise would be immoral.

Utilitarianism is believed by some scholars to be a thinly disguised justification for capitalistic excess, which they fear will start mankind down a slippery slope to merchandizing human beings. Utilitarians like John Stuart Mill were dead set against such slavery, which they believed could hardly maximize happiness—but scoundrels have been known to take refuge behind utilitarian arguments. A more serious problem most religions have with utilitarianism is that it only counts sentient creatures—if something can feel neither happiness nor pain, it is simply not a part of the equation. It is a "nonmoral" agent, bereft of human rights.

Utilitarianism is a laudable attempt to quantify and calculate units of pleasure and pain, but it is far from an exact science. Even in ordinary cases, it poses some difficult questions. In extraordinary cases, like the quagmire of embryonic stem cell research, it is merely a starting point for the discussion.

Stem cell ethics committees generally consider several overlapping concerns when they review a research project:

- The laws governing human embryo research.
- The scientific goals and techniques of the research.
- Potential applications of the research.
- The procedures by which human tissues are obtained.
- Federal funding restrictions regarding human embryo research.
- Alternative methods that might achieve the same goals.
- The value of scientific freedom.
- Public opinion.
- The recommendations of other committees.

Ethics committees often engage religious scholars, with the understanding that multiple views should be expressed. However, certain of these spiritually oriented thinkers, although well versed in religious philosophy, have little special education in modern reproductive and medical therapies. Some ethicists believe that while including the views of this segment of society is important, it is not wise to give them a formal position because it institutionalizes the very conflict of values that church-state separation was designed to solve.

Others, such as Robert Veatch, believe that expert groups cannot be representative of a modern pluralistic society. While this is usually true, it ignores an important problem: Without a full presentation of the facts, how are people expected to make informed decisions? And, after the facts have been delivered, shouldn't one hope that at least some expertise has been gained?

Ethics committees have been blasted for being biased, for having too many religious people or too few, for being beholden to a corporate philosophy, and for many other perceived or real sins. However, the mere existence of these committees is a positive sign that researchers, governments, and corporations are all looking to do the right thing.

The Scientific Debate

Most of us in the field think the push to define adult or embryonic cells as better than the other is a lot of folly.

George Daley
Whitehead Institute
MIT

If scientists could create embryonic stem cells without using an egg, they would. However, at this time, other ways of creating ES cells are not known. As described in the preceding chapters, before a woman's eggs are fertilized, they each have their own identity, defined by the unique genetic mixtures of maternal and paternal genes that occur during meiosis (Chapter 3). The woman herself was once a tiny egg in her mother's ovary, a unique mixture of her maternal grandparent's genes. And so on before that, in an unbroken chain, back through each maternal egg to the first cell on the face of the earth, several billion years ago. Like the Buddhists, biologists share a fondness for continuity in life.

The moment when a sperm and an egg meet provides a romantic starting point for life, but in reality that moment is pretty fuzzy. As described in Chapter 5, there are dozens of steps involved in fertilization, but at that first meeting there is no blending of male and female genes. Therefore, if combining DNA is a signal event, it can't be said to happen at the moment of fertilization. Therefore, from the biological point of view, conception fails as a signpost for novel life.

It is also clear that not every zygote is destined by nature to complete the journey. Along the way, problems may occur that cause up to two thirds of all zygotes to be lost. Some religions prohibit abortion because they believe it is God's plan that each conception should lead to a baby. However, the natural path from conception to birth is not so assured. Until about day 14, the blastocyst has the potential to split and form twins. The idea of personhood before this time is counterintuitive because we are not accustomed, mentally or legally, to seeing people split in two. If the original blastocyst had full human rights, could it sue its cloned twin for stealing half of its "body?" Until the potential of becoming a unique human is actually manifest in the embryo, we can't confer humanhood without provoking paradoxes.

The 14-day-old embryo can split because it has no organized tissue to dismantle. There is no hint of a nervous system yet, either centralized or peripheral. For these reasons, several governments, including England's, allow research on human embryos up until the second week, when the "primitive streak" is formed (Chapter 9).

Scholars since Saint Augustine have assumed that sentience is delayed in the developing embryo. Biology has confirmed these assumptions: the neural tube forms between weeks 3 and 4 of embryonic development (Chapter 9), and by week 8 the brain starts to take form. By the third trimester, brain waves can be recorded. Might sentience be a useful marker for the start of life? This would bookend well with the end of life as it is viewed by most religions. Consider the following:

When organ and tissue transplants were first proposed in the 1950s, all major religions came out forcefully against them. Transplants from dead people into the living were something straight out of Frankenstein. But within a few years—as the prognosis for transplant recipients improved—attitudes softened. Today, all major religions, with the notable exception of Shinto, have accepted organ transplants as a moral therapy. One of the things they insist on is that the donor be dead before essential organs are harvested. Death is defined as flat-line brain waves. In other words, the major religions consider that humanhood ceases upon brain death. This seems like a reasonable dividing line; we instinctively know that without a brain, the person is gone.

If mental function is essential to humanhood, and its extinction sufficient to prove "nonhumanhood," then it might be reasonable to assume that humanhood also starts with mental function. This provides a bright line—that point in development where brains are developed and senses are awakened—that could determine when life begins. This has the evidence of biology to back it up and is buttressed by the appeal of common sense. According to polls, this is what the majority of Americans believe.

A vague consensus, however, is no reason for scientists to be cavalier. They know they are studying life, and that life is sacred. Undeniably, scientists are generally less squeamish when it comes to exploring life than the general population. They know how important it is to learn and observe from nature. They know that diseases cannot be cured without observation. It is clear, therefore, that dead embryos will need to be dissected to discover how to cure birth defects. The lay person may not want to know all the details, but they are generally grateful for the curative therapies that result.

The public has been led to believe that there is a scientific conflict between research on adult and embryonic stem cells. This is important because if adult stem cells, which don't carry so much ethical baggage, could work, then there would be no need to study ES cells. However, most scientists don't frame the debate in those terms, and most researchers insist that both lines of research are needed.

Neil Theise, an adult-stem cell researcher at the New York University Medical School, says, "Losing the opportunity to study embryonic stem cells could mean that we could lose out on some of the enormous therapeutic potential offered by these cells."

Catherine Verfaillie, who has made several discoveries about adult stem cells says, "My message has always been, even though we're excited about the adult cells, it's too early to say that they will replace embryonic stem cells." She points out that much of her research has depended on discoveries made in the embryonic stem cell field and she recommends continuing research on both.

In September 2002, Irving Weissman and Amy Wagers used fluorescent tags to show that adult stem cells can be converted to blood stem cells, but they found virtually no other stem cell products. Wagers concluded that adult stem cell plasticity is not "as robust as it is claimed to be." Other work by Weissman and Wagers has shown that hematopoietic stem cells circulate through the various tissues of the body, which may lead to their misidentification as organ-specific stem cells. These results still need to be reconciled with other experiments that demonstrate greater plasticity of adult stem cells.

The rules of science demand objectivity, which ultimately weeds out religious or political agendas. Scientists will decide which cells to use for which therapies based on experimental findings, but only when those experiments are allowed.

The Future of the Debate

Science faces—in ethics and religion—its most interesting and possibly humbling challenge …while religion must somehow incorporate the discoveries of science to retain credibility.

Edmund O. Wilson
Consilience

As James Thomson can attest, each advance in cloning and stem cell technology has been greeted by howls of protest. The usual concern is that the scientists are trying to play God, but are instead creating monsters using dark and mysterious technologies. Unfortunately, the language of the stem cell investigator is still baffling to the majority of citizens. Instead of scientists valiantly struggling to cure disease, many see Dr. Frankenstein tending his embryo farm. Scientists, having made their research inaccessible to the layperson, must shoulder some of the blame for the confusion.

Scientific ignorance is the driving engine of the antiembryonic stem cell movement. Polls show that the less education people have, the less likely they are to support ES cell research. As people learn more about the biology—and after the first cures get reported—attitudes may soften.

However, to date, opposition to human ES cell research in the United States has been effective. When the American Heart Association offered to support and even fund certain stem cell projects, a mail-in campaign organized by the Roman Catholic

Archbishop of St. Louis compelled the AHA to rescind their offer. The same sequence of initial support and then abrupt disavowal occurred with the American Cancer Society and the American Red Cross after similar write-in campaigns.

The result is that the research has been driven into the private sector and overseas. Companies in the private sector depend on patent royalties and trade secrets, and they're not always eager to publicize the results of their research. The ironic upshot is that the research is being driven back into the dark, where there is no federal oversight, no standardization, and no data sharing. This is probably not what most of the letter writers were hoping for. Religion, for all its passion, cannot decide the issue for us. There are simply too many diametrically opposed theologies—many claiming to be the actual word of God—to reach a consensus.

Fortunately, in democratic societies, these decisions are up to the citizens and their Constitutions. Unfortunately, many of the people feel ill equipped to render an informed decision. Science is hard to understand, religion offers conflicting advice, the law is a muddle, and ethics are too abstract. In the face of all that, about two thirds of Americans polled still favor embryonic stem cell research.

The answer, ultimately, may come down to "common sense," which will dictate to most citizens that a dot of cells is not a human being, and that it is wrong to let people suffer when there may be a cure at hand. This prediction is based on the long history of scientific progress and its attendant backlashes. Once a science yields fruit and starts to improve the lot of the average person, the backlash melts away.

That said, a smarter, more compassionate plan would be to create a War on Cancer approach to speed the development of embryonic stem cell research. Its funding should be in line with its incredible promise. At the same time, strict protocols should be established to ensure that babies are not being conceived merely as fodder for research, and that experiments are vetted by ethics committees. The subject is too important to be left unobserved and unregulated.

Today's federally funded scientists are dependent upon peer-reviewed grants and publications. Their work is scrutinized by their peers, the public, the press, the government, and the church. There is no place to hide. By defining areas of research that have been cleared of moral minefields, guidelines can actually be liberating to researchers. The uncertainty of the current regulatory environment is discouraging investment, both economic and educational, in this profoundly important science.

The people with the most to gain are those patients who are sick or dying from one of the diseases targeted by stem cell therapy. For them the long-windedness of the debate is often unbearable. The public would be best served by the approval of federal funding, which will help scientists learn more about the miracles of embryogenesis and fertility, while at the same time providing possible cures for heart disease, Parkinson's, arthritis, Alzheimer's, diabetes, and more. In the bargain, federal funding will provide stringent guidelines for the research, and ensure that the government-funded results get the widest possible exposure.

For the sake of the millions who could benefit from stem cell therapy, this issue demands resolution—and quickly.

Additional Readings

Green, R. M. (2001). *The Human Embryo Research Debates.* New York: Oxford University Press.

Green, R. M., DeVries, K. P., Bernstein, J., Goodman, K. W., Kaufmann, R., Kiessling, A. A., Levin, S. R., Moss, S. L., and Tauer, C. A. (2002). Overseeing Research on Therapeutic Cloning: A Private Ethics Board Responds to Its Critics. *Hastings Center Report* 32: 27–33.

Holland, S., Lebacqz, K., and Zoloth, L. (2002). *The Human Embryonic Stem Cell Debate.* Cambridge, MA: The MIT Press.

Glossary

abembryonic: The trophoblast cells directly opposite the inner cell mass of a blastocyst.

acetylcholine: A neurotransmitter in vertebrate nervous systems found both in the brain and in peripheral nerves.

allele: The gene on each of two chromosomes in a pair, one inherited from the father and one from the mother; different mutational states in the same gene within the organism and/or within the gene pool for the species.

allogeneic: Pertaining to the same gene pool, but not the same organism.

Alzheimer's disease: A disease involving death or injury of brain cells characterized by memory loss and dementia.

amnion: The membranes arising from polar trophoblast cells that completely encase the developing fetus.

androgenote: A zygote, and the resulting cells that arise following cleavage, that has been manipulated to contain two male pronuclei instead of one male and one female pronucleus.

anemia: A medical condition characterized by abnormally low levels of hemoglobin in circulating blood.

aneuploid: Possessing a chromosome number that is not characteristic of the species

aplastic anemia: A medical condition characterized by low levels of hemoglobin in circulating blood resulting from decreased production of red blood cells.

assisted reproduction: Procedures performed by health care providers to increase the possibility of pregnancy.

assisted reproductive technology: The medical and laboratory procedures developed to increase the possibility of pregnancy through manipulations of eggs and sperm in laboratories.

astrocytes: Star-shaped glial cells that are numerous in the brain and the spinal cord that regulate the extracellular environment; they also respond to injury and may be a type of adult neural stem cell.

autocrine: The process whereby a cell produces a substance that brings about stimulation of one of its own receptors; the usual circumstance involves secretion by the cell of a hormone or growth factor that then binds to a receptor on the cell's own plasma membrane.

bilaminar germ disc: An advanced inner cell mass stage of embryonic development that appears around the time of implantation; it is characterized by a layer of endodermal cells adjacent to the blastocoel and a layer of ectodermal cells adjacent to the developing amniotic cavity.

blastocoel: The fluid-filled space inside the blastocyst.

blastocyst: A ball of trophoblast cells sealed together with tight junctions that contains the inner cell mass of undifferentiated embryonic stem cells that gives rise to the embryo.

blastomere: A large, undifferentiated cell resulting from cleavage of an activated egg.

bone marrow: The population of blood and immune stem cells that resides in the spongy areas inside bones, especially long bones.

bone marrow transplantation: The medical procedure involving transplantation of cells from the bone marrow to a recipient with the goal of regenerating healthy blood and immune systems within the recipient. Bone marrow may be transplanted as an allograft from a donor or autologously from the patient.

cAMP: Cyclic adenosine monophosphate, a common intracellular second messenger produced by enzymes that reside in or near the plasma membrane of the cell in response to binding of a growth factor or hormone to the cell membrane; a common target for cAMP is activation of protein kinase A.

cardiomyocyte: A general term for cells contributing to heart muscle.

caudal: A position near the tail or hind part of the body.

cell: The smallest living unit of mammals and most plants and animals. Cells are defined by an outside (plasma) membrane and contain a full complement of genes characteristic of the species plus a mixture of proteins and other molecules.

cell differentiation: The transition of a cell to the state of gene expression that allows it to carry out the functions specifically required of its tissue or organ of residence; differentiation is frequently accompanied by changes in the appearance of the cell.

cell division: A process that includes replication of all the DNA in each chromosome to create a second set of chromosomes that are then divided equally into the two halves of the cell, which pinches the plasma membrane in the middle to create two new, identical daughter cells.

cell fusion: The process of two cells coming together through fusion of their plasma membranes.

cephalic: Pertaining to the head or brain.

chimera: An organism containing two or more genetically distinct cells. An example is a mixture of blastomeres from two mouse embryos that form one blastocyst, which gives rise to a mouse with tissues that contain daughter cells from both mouse embryos.

chorionic villous membrane: A trophoblast cell derivative that forms the interface between the developing embryo and the lining of the uterus.

chromosome: A collection of genes contained within one long, double strand of DNA, which condenses into a distinct entity during cell division.

chromosome pairing: The process of alignment of pairs of matching chromosomes in such a way that one chromosome from each pair will divide with each daughter cell.

clonal expansion: The repeated division of a single cell into a large population of identical daughter cells.

clone: Greek for twig, which can be stripped from a tree and planted to grow an identical tree; used as a scientific term to mean genetically identical.

cloning: Cloning is the act of creating a clone, either as single cells or as individuals. For example, twins are clones of each other, and a cow arising from nuclear transplantation into an unfertilized egg is a clone of the donor of the nucleus.

compaction: The process of blastomeres forming bonds between their plasma membranes that results in adherence of the cells together.

constitutive expression: The continual expression of a gene into its protein product without the need for specific, outside stimulus.

corona radiata: The term used to describe the microscopic appearance of a layer of cells next to the matured egg that are spread out from touching each other by the synthesis of the glycoprotein hyaluronic acid.

cumulus cells: The cells surrounding the maturing egg within the developing follicle in the ovary.

cytoplasm: The collection of proteins, carbohydrates, and other molecules and structures that constitute the cell outside of the nucleus.

cytoplast: A cell after removal of its nucleus.

cytostatic factor: The term used to describe the proteins that bring about arrest of the cell cycle; for eggs in metaphase II arrest, the protein cMos is a component of cytostatic factor.

dedifferentiation: The process of losing specialized cell functions and gaining the capacity to divide and give rise to a wide range of alternate cell types.

delayed implantation: A delay of the process of implantation into the lining of the uterus past the normal period of time following fertilization; many mammals have delayed implantation of eggs fertilized during the time they are producing milk for their young; during the delay, the fertilized egg may develop to the blastocyst stage and then enter a period of relative quiescence until lactation ceases, at which time implantation proceeds normally.

diabetes: A disease characterized by the body's inability to properly metabolize blood sugars. Type I diabetes is caused by the loss of insulin-producing cells in the pancreas. Type II diabetes is more common and results from a wider range of cellular disorders.

differentiation: The typical course of embryogenesis involves the specialization of stem cells into progenitor and precursor cells and finally into adult tissue through the process of differentiation, or specialization.

diploid: Containing both copies of chromosomes of each pair, generally one from the mother and one from the father.

dizygotic twins: Twins arising from two fertilized eggs; also called fraternal twins.

dopamine: A catecholamine neurotransmitter largely localized to cells in the substantia nigra and striatum. The death of these cells is the cause of Parkinson's disease.

dorsal: A position near the back (top) of an organism.

dyskinesia: Difficulty in controlling movement; drugs used in the treatment of some diseases like Parkinson's may cause dyskinesia as a side-effect.

electrofusion: Membrane fusion caused by a small jolt of electricity.

embryo: The developmental stage that follows implantation and precedes organogenesis.

embryonal carcinoma: A malignant tumor arising from embryonic cells, usually germ cells.

embryonic diapause: The relatively quiescent period of embryonic development that accompanies an hormonally induced biological delay in implantation; an example is the three-week delay in implantation associated with lactation in a mouse.

embryonic stem cells: Undifferentiated cells derived from the inner cell mass of a blastocyst that will give rise to every cell of the adult organism; in laboratory culture, embryonic stem cells can be maintained in an undifferentiated, dividing state indefinitely.

endogenous retroviruses: Genetic copies of retroviruses that are thought to be due to ancient infections that have become a permanent part of the organism's genome; under certain conditions in the cell, the virus genes may be turned on and produce virus particles.

endometrium: The lining of the uterus that thickens in preparation for pregnancy.

endoreduplication: The replication of chromosomes without cell division.

enucleated: The removal of the nucleus from a cell; the term is also applied to removal of chromosomes in a metaphase plate.

enzyme: A protein that enables chemical reactions to occur at body temperatures.

epiblast: The ectodermal layer of inner cell mass.

epidermal growth factor: One of the first growth factors to be discovered, it acts on a variety of cell types besides epithelial, including neurons.

epididymis: The highly convoluted tubular organ that collects and stores spermatozoa from the seminiferous tubules of the testis.

ES cells: Embryonic stem cells, typically cultivated from the inner cell mass of a blastocyst.

estrogen: A steroid hormone produced from testosterone by specialized cells in the ovary and some other tissues.

ethics: A system of values and rules of conduct that attempts to discriminate between good and bad behavior.

eugenics: "True" genes, a theory calling for the use of scientific principles to select toward a given set of desired traits.

feeder layer: A layer of cells in the culture dish that is living but has been treated (e.g., exposed to X-rays to damage the DNA) to prevent cell division.

fertilization: The process of sperm entry into an egg followed by formation of individual pronuclei, DNA replication, and cell division to create two new cells with new chromosome combinations.

fetal stem cells: The progenitor cells giving rise to the tissues and organs of a fetus.

fetus: The developmental stage of growth following organogenesis.

FGF4: Fibroblast growth factor number four, discovered to be an important growth factor for stem cells.

fibroblast: Progenitor cells derived from the mesoderm; fibroblast cells derived from mice are used as a feeder layer for stem cells in laboratory culture.

follicle: The structure within the ovary that houses the maturing egg; at the beginning of the monthly menstrual cycle, cells associated with the egg are stimulated to divide and secrete estrogen, which further stimulates their division and the accumulation of fluid, thus creating a blister-like structure about 2 centimeters in diameter at the time of release of the egg at mid-cycle.

follicle stimulating hormone: The protein hormone synthesized and secreted by the pituitary gland of men and women that binds to receptors in both the testis and ovary; classified as a gonadotropin.

FSH: Follicle stimulating hormone.

galactosidase: An enzyme that hydrolyzes sugars that contain galactose; the gene that encodes it, *lac*Z, is a bacterial gene that is commonly used as a reporter gene in mammalian cells.

gene: A specific combination of deoxynucleotides that encodes either messenger RNA, ribosomal RNA, or transfer RNA.

gene pool: A term that refers to the collection of all genes in a given population.

gene replacement: Replacing the gene within the cell with a new gene or partial gene sequence.

genome: Refers to all the genetic information within an organism.

germ cells: Eggs and sperm and their progenitor cells.

germinal vesicle: The nucleus of a growing oocyte.

gestation: The period of embryonic and fetal development within the uterus.

golgi apparatus: The membrane structure responsible for protein processing and transport throughout the cell.

gonadal ridge: The region in the posterior mesoderm that accumulates germ cells migrating from the yolk sac.

gonadotropin releasing hormone: A small protein synthesized and secreted by the hypothalamus that stimulates the pituitary to release gonadotropins.

gonadotropins: Hormones that stimulate the ovary and testis.

gonads: General term for both male and female germ-cell-producing organs and their embryologic precursors.

granulosa cells: Steroid-hormone-producing cells that surround the oocyte and line the growing follicle within the ovary.

growth factors: General term for a large family of proteins or other molecules that stimulate receptors and intracellular pathways that ultimately lead to cell growth and division; may also have other effects on cells.

growth hormone: A small protein secreted by the pituitary that stimulates many tissues and cells by binding to specific receptors, most notably at the ends of long bones in the legs and arms.

growth hormone releasing hormone: A small protein secreted by the hypothalamus that causes release of growth hormone from the pituitary.

gynegenote: The zygote and blastomeres that arise following cleavage of an egg that has been manipulated to contain two female pronuclei.

haploid: Containing only one copy of each chromosome of the species.

hatching: The process of blastocyst escape from the zona pellucida.

hemangioblast: A blood cell progenitor that appears early in embryogenesis.

hemophilia: A disease characterized by absence or malfunction of one of the many factors involved in forming a blood clot; affects mostly men.

hepatic: Pertaining to the liver.

heteroploidy: The state of containing other than the normal number of chromosomes for the cell.

hFES cells: Human embryonic stem cell derived from the inner cell mass of a blastocyst formed following fertilization of an egg by sperm.

homolog: Refers to gene sequences or proteins that share a common evolutionary or ancestral origin.

homologous chromosome: Refers to one of the chromosomes of a pair, each of which was derived from a different parent.

homologous recombination: Genetic substitution that results from the binding of specific DNA sequences at the ends of newly introduced DNA that leads to replacement of the original DNA in a chromosome with the new DNA.

hormone: A substance, generally a protein or a steroid, that is synthesized and secreted by body tissues, usually referred to as glands, that can elicit a specific action from nearby cells or distant cells expressing the appropriate receptor.

hPS: Acronym for human stem cells derived from parthenogenetic activation of an unfertilized egg.

human cloning: Refers to the creation of a new human being genetically identical to an existing human by the same methods used to create cloned animals; also sometimes used to describe the creation of laboratory lines of identical human cells.

human immunodeficiency virus (HIV): A retrovirus that is responsible for acquired immunodeficiency syndrome (AIDS); HIV enters cells through receptors meant to elicit immune response, synthesizes a DNA copy of its RNA genome that can integrate and becomes a permanent part of the cell's chromosomes.

hyperstimulation: A condition of overresponse to a stimulus such as a hormone.

hypothalamus: An area of the brain beneath the thalamus that carries out several functions to control metabolism including stimulating release of hormones from the pituitary.

ICM: inner cell mass, the group of cells within the blastocoel of the blastocyst.

immune: Protected against infection by specific responses from the immune system.

immune response: The production of antibodies and other molecules of defense by the blood cells and lymphatic cells that constitute the immune system of an organism.

immune system: Comprising many cell types and structures, including white blood cells and tissue macrophages, most of which are generated by hematopoietic stem cells in the bone marrow.

immunodeficiency: The condition of being incapable of mounting a normal, effective immune response against a pathogen.

implantation: The process of attachment by an early embryo, usually at the blastocyst stage, to the endometrial cells lining the uterus followed by immediate outgrowth of embryonic trophoblast cells between the endometrial cells.

imprinting: The process of influencing gene expression without altering the nucleotide sequence within the gene; termed epigenetic control on gene expression, imprinting occurs differently within the testis and ovary.

inheritance: The process of passing characteristics from parents to progeny.

insulin: A peptide hormone secreted by the pancreas and used by the body to process sugars. A lack of insulin resulting from defects in pancreatic cells is the cause of type 1 diabetes.

integration: The process of insertion of new DNA sequences into the DNA strands that comprise the chromosomes.

interphase: The phase of the cell cycle in which the nucleus is visible.

IRB: Internal (or investigative) review board; common acronym for the group of professionals that review research protocols that involve human subjects.

islets of Langerhans: small groups of cells in the pancreas that produce and release insulin.

IVF: A common acronym for the assisted reproduction technology, in vitro fertilization, which involves recovery of eggs from ovaries and combining them with sperm in laboratory culture to promote fertilization and egg cleavage.

karyoplast: Term used to describe a nucleus that has been removed from the cytoplasm of its cell.

kinase: An enzyme that adds phosphate groups to amino acid side chains in a protein.

lactate: The production of milk by mammary glands; the ester of lactic acid.

lactational delay: The suspension of embryonic development during the period of milk production.

lacZ: The gene that produces beta-galactosidase.

luteinizing hormone: A protein hormone synthesized and secreted by the pituitary gland in the brain that stimulates both the ovary and the testis; a gonadotropin.

LH: Luteinizing hormone.

MAP kinase: The enzyme that adds phosphate groups to mitosis activating protein, which is itself also a kinase.

MAPs: A new term, mature adult progenitor, used to describe stem cells from adult bone marrow with the capacity to differentiate into a wide range of cell types.

maturation promoting factor: The term used to describe the cytoplasmic factors responsible for bringing about the resumption of meiosis in frog oocytes; comprising several cell factors, including CDC/cyclin complexes.

meiosis: The process of dividing chromosomes twice without an intervening cycle of DNA replication and chromosome duplication; results in cells with a haploid number of chromosomes.

mesenchyme: A descriptive term for cells and extracellular connective proteins that constitute the unspecified areas of the developing embryo and fetus.

mesoderm: The middle layer of cells of the three germ layers of a developing embryo. It gives rise to blood, muscle, and skeletal tissue.

metaphase: The stage in the cycle of a dividing cell at which chromosomes have fully condensed and become attached to the proteins that constitute the mitotic spindle but have not begun the process of dividing apart into alternate halves of the cell.

metaphase plate: The imaginary plane in which chromosomes are positioned at metaphase.

micromanipulators: Instruments associated with microscopes that are capable of holding tools, usually glass, and adjusting their position within dimensions smaller than individual cells.

micron: A micrometer; one thousandth of a millimeter, approximately 1/25,000 of an inch.

micropipettes: Hollow tools held by micromanipulators that are capable of delivering or recovering microliter amounts of substances, cell fractions, or cell organelles.

mitochondria: The self-replicating, membrane-bound structures within the cell that generate adenosine triphosphate (ATP), which is the principal source of energy for the cell; also contain their own DNA that encodes for some mitochondrial-specific proteins and have been used as an evolutionary tool to prove that most of the mitochondria in an organism derive from the egg, not the sperm.

mitosis: The process of dividing chromosomes, cytoplasm, and cell membranes into two equal daughter cells.

monozygotic twins: Two offspring arising from one fertilized egg.

myelin: A specialized, low-conductance cell membrane wrapped around vertebrate axons that serves to insulate and protect nerve impulses; produced by glial-type cells termed oligodendrocytes in the central nervous system and Schwann cells in the peripheral nervous system.

nerve growth factor: A peptide that promotes the growth of nerve tissue. NGF can be used in vitro to encourage the differentiation of nerve cells from embryonic stem cells.

neuron: A nerve cell of the brain or spinal cord that conducts chemoelectric signals.

neurospheres: The ball of cells arising in laboratory culture in the presence of nerve growth factor that consists of neurons, astrocytes, and other glial cells clumped together.

neurotransmitters: The small molecules released by neurons at a gap called a synapse that connects to the target nerve cell; different types of nerve cells secrete different neurotransmitters, including acetylcholine, adrenaline, serotonin, and dopamine.

nestin: An intracellular protein expressed by neural cells.

nidation: The process of communication between the early embryo and the uterine cells, attachment to the uterine cells, invasion of the embryonic trophoblast cells, and the early formation of the placenta.

notochord: The embryonic structure that defines the spinal cord region of the developing fetus.

nuclear membrane: The membrane that separates the nucleus from the cell cytoplasm.

nuclear transplantation: The transfer of a nucleus from one cell to another; also frequently used to describe the fusion of a somatic cell with an egg, which results in transfer of both the somatic cell nucleus, cytoplasm, and plasma membrane.

nuclear transplantation stem cells: The pluripotent cells arising from blastomeres stimulated to divide following transfer of a somatic cell nucleus, or fusion of the somatic cell, into an egg cytoplast that is then activated to initiate DNA replication and cell division.

nucleolus: The nuclear region for transcription of ribosomal RNAs and assembly of ribosomal subunits.

nucleus: The structure within the cell that contains all the chromosomes and provides the membrane scaffolding required for DNA replication and transcription.

oligodendrocyte: A type of glial cell in the brain and spinal cord that sends processes out to nerve cells and wraps around and coats the axon in a "myelin sheath," which insulates the axon, permitting low-loss electrical conduction along its entire length.

oocyte: General term to describe all the stages of egg cell growth and maturation.

oogonia: Egg progenitors in the fetus that divide to provide a population of eggs for the ovary.

organ: A collection of specialized cells and tissues that carry out a defined function within the body, such as the heart.

organ transplantation: The surgical insertion of a new organ to replace, or assist, an existing organ; new kidneys are frequently inserted into the blood supply of an existing kidney without removing the defective kidney.

organism: Any living thing, plant or animal, single or multicelled.

ovarian follicle: The blisterlike structure within the ovary that enlarges in response to the hormone signals that stimulate cell division by the granulosa cells that surround and support the final stages of maturation by an egg.

ovasome: A new term coined to describe the use of eggs to create somatic cells for stem cell therapy rather than new offspring.

ovulation: The process of final maturation and release of an egg from the ovary.

pancreas: The organ that secretes insulin in response to blood sugar.

Parkinson's disease: A disease of the loss of function of dopamine-producing nerve cells located in the midbrain that results in limb tremor and slow, stiff motion.

parthenote: The blastomeres and cells that arise following egg activation in the absence of sperm.

pathogen: Any microorganism, such as a bacterium or virus, that causes disease.

perinatal: Pertaining to the few weeks of time before and after birth

perivitelline space: The space between the plasma membrane of the egg and the inner aspect of the zona pellucida that surrounds the egg.

phosphatase: An enzyme that removes phosphate groups from proteins.

pituitary: The gland in the brain near the hypothalamus that synthesizes and secrets several protein hormones, including growth hormone and the gonadotropins, follicle stimulating hormone, and luteinizing hormone.

plasma membrane: The membrane that encloses and separates the cell from the outside environment.

pluripotent: Cells that have the ability to differentiate into more than one type of tissue—like embryonic stems cells—are pluripotent. ES cells, because they can't form trophoblast tissue, are not considered to be totipotent, or capable of producing an entire organism.

polar body: The small cell that results from the asymmetric division of the egg; the first polar body contains the diploid set of chromosomes resulting from the first meiotic division; the second polar body contains the haploid set of chromosomes resulting from the second meiotic division.

polar trophectoderm: The layer of outside cells that overlies the inner cell mass cells in the blastocyst.

polyadenylation: The protein synthesis machinery within the cell depends on a series of adenosine molecules at the end of a messenger RNA to stimulate translation of the message into protein; although there are exceptions, in general, messenger RNAs that are not polyadenylated are in a state of storage and are not expressed as protein.

polygamy: The social practice of marrying more than one person.

polyploid: A cell containing more than one pair of haploid chromosomes.

precursor cell: A cell, typically derived from a progenitor cell, that is fated to differentiate into one or two specific cell types.

pre-implantation: The period of development of an organism that precedes implantation in the uterine lining.

primary oocyte: The stage of oocyte maturation that includes a not fully grown egg arrested in the prophase of meiosis and surrounded by one layer of granulosa cells.

primitive ectoderm: The layer of inner cell mass cells immediately under the polar trophectoderm of a newly implanted blastocyst that is forming a bilaminar embryo.

primitive endoderm: The layer of cells that lines the inner cell mass and the blasto-coel cavity on the inner aspect of the blastocyst.

primitive node: The pit that develops on the dorsal side of the bilaminar embryo as a result of outgrowth and expansion of mesodermal cells between the primitive ectoderm and the primitive endoderm resulting in a trilaminar embryo.

primitive streak: The groove that develops from the primitive node to the outer edge of the embryo at the transition from a bilaminar to a trilaminar stage of development; the appearance of the primitive streak defines the embryo with respect to dorsal, ventral, caudal, and cephalic regions.

progenitor cell: Multipotent cells that can differentiate into a limited set of cell types.

pronucleus: The large nucleus with a few, prominent nucleoli, formed by an activated egg to contain either egg chromosomes or the sperm head, if fertilized, or formed by remodeling a somatic cell nucleus transplanted into the egg.

prophase of meiosis: The period between replication of the complete set of chromosomes and the onset of meiosis during which chromosomal crossing-over occurs; in most mammals, including humans; this period extends from fetal development to puberty.

puberty: The onset of sperm production in males and oocyte maturation and release in females.

quickening: The time during a pregnancy when the fetus has grown to a size sufficient for the mother to feel its movements within the uterus.

reporter gene: A well-characterized, easily detectable gene that is attached to a gene of interest; because of their close association, if the reporter gene is expressed, the gene under study is also assumed to be expressed.

retinal degeneration: A disease process that results in loss of retinal cells in the eye and consequent loss of vision.

retinoic acid: A small molecule related to vitamin A that plays a major role in guiding development of early embryonic structures.

retrovirus: A virus whose genome is made of RNA and replicates by causing the synthesis of a DNA copy of its RNA genome that then serves as template for new RNA synthesis for progeny virus.

ribosomes: The two-subunit, protein-RNA structure responsible for organizing all the intracellular elements necessary to compile the specific amino acid sequences of proteins as directed by the nucleic acid sequences in the messenger RNA.

"S" phase: The DNA synthesis phase of a cell cycle.

second messenger: A term applied to a wide variety of intracellular molecules that promote an action by the cell in response to binding of extracellular molecules, such as hormones or growth factors, to receptors in the plasma membrane.

somatic cell: A body cell, as opposed to a reproductive (germ) cell like an egg or a sperm.

sperm: The male germ cell.

spermatogenesis: The process of creating new sperm within the seminiferous tubules of the testis.

spermatogonia: The progenitor sperm cell that continually divides to produce new sperm.

stem cell: An uncommitted cell that can divide continually without loss or gain of chromosomes or genes and gives rise to every cell type in the body.

syngeneic: Arising from the same or genetically identical organism.

syrinx: A fluid-filled swelling in the spinal cord that accompanies spinal injury; may prevent nerve regeneration at the site and may produce pain and weakness.

tetraploid: Four copies of each chromosome.

therapeutics: A substance used to correct a disease state.

tissue: A collection of cells coordinated to perform a specific body task.

tissue transplantation: The process of transferring cells, such as bone marrow cells, from one person to another to correct a disease in the recipient.

transcription: The process of creating an RNA copy of a DNA sequence in a gene.

transfection: A process that inserts new DNA into a cell.

transgenic: An organism with foreign DNA in its cells that was introduced at or near the time of fertilization.

triplet codon: The group of three nucleotides in a messenger RNA that determines which amino acid to add to the growing protein within the ribosomal complex; more than one codon can define each amino acid, and some codons signal the beginning and ending of the protein.

trophoblast: The outside cells of a blastocyst that will give rise to the placenta and contribute to other extra-embryonic membranes such as the amnion.

ultrasound waves: Sound waves with a wavelength too short to be heard by the human ear.

ventral: A position near the belly of the body or the anterior of an object.

von Wildebrand's disease: A disease of the blood clotting system that involves defects in platelet aggregation and affects both men and women.

zona pellucida: The glycoprotein coat synthesized by the egg that surrounds it and protects it during the pre-implantation period of development.

Index